INSPIRATIONAL CONCEPTS IN THE HIGH SCIENCES

INSPIRATIONAL CONCEPTS IN THE HIGH SCIENCES

Cedric Paul Harriott (Heh Heru)

2015

Copyright © 2015 by Cedric Paul Harriott (Heh Heru)

All rights reserved. This book or any portion thereof may not be reproduced or used in any manner whatsoever without the express written permission of the publisher except for the use of brief quotations in a book review or scholarly journal.

First Printing: 2015
Third Edition

ISBN: **978-1-951881-00-9**

Cultured Melanin Literary & Visual Arts Studios
44 Shipping Place
PO Box #4042
Baltimore, MD 21222

https://www.culturedmelanin.studio

Original Cover Photograph By: Sergey Dryutskiy
Email: graphicshotgallery@gmail.ru
Instagram: https://www.instagram.com/graphicshot

DEDICATION

To whatever creative forces brought my spirit essence to the human plane to experience this form.

To my family and the transitioned ancestors that reared me, and guided my mind in pure form.

To my ancestors in Africa that I was robbed from knowing but have since begun to find.

To the many friends now lost, and to those now manifested.

To the ones that love me and to those that loathe me.

To the other aspects of life out there in the cosmos.

To the sentient beings that I have committed my life to protecting and experiencing life alongside.

To the many musicians and artists that have left their treasures for the world's pleasure & inspiration.

To all of the thousands of scientists and philosophers that have dedicated countless nights, and many headaches to the enigmatic questions of life.

Thank you, for without your criticism, tremendous support, inspiration, conversations, sparring sessions or abandonment respectively, I would have never achieved my wakefullness.

Contents

Acknowledgements i
Preface iii
Introduction 1

Essence of Numerology: 9
   Authors Note for Essence of Numerology: 11
   From the Everything of Nothing – 0 13
   And Then There Was One: 1 17
   The Masculine & the Feminine; Duality: 2 19
   Trinities, Triangles, Third Eyes and the Child: 3 23
   The Operative Builder; the Masonic Square: 4 27
   Mind & the Womb: 5 31
   Carbon + Death? + Life + Transcendence: 6 35
   Completion & the Seventh Heaven: 7 39
   The Voided: 8 41
   Oh: 9 43
   Kemetic (Egyptian) Numerology 45
   Let Us Break Bread: 0-9 47

Essence of Mathematics: 51
   Pre-Fibonacci & Thee Fibonacci Sequence 53
   The Golden Rule / The Golden Ratio: 1.618 58
   The 9 Code; Its Sublime Insertion into Angles 66
   Sacred Geometry: The Language of the Creator 71

Essence of Astronomy + Astrology: 75
   Hexagonal Storm on Saturn 77
   Observational Universe 81
   Experiment: The How Far Thought Projection 83
   Planetary Sounds 93
   Zodiacs, Constellations, Ages & Cycles in the Heavens 95
   Great Red Spot; Jupiter 103
   Black Holes 107
   Quasars 113

- The Dogon Tribe of West Mali 119

Essence of Magnetism: 123
- Earth Magnetics 129
- Local Magnetism 133
- The First, the Heart. 135
- Copper Pipe & A Magnet Experiment 137

Essence of Music & Audiology: 141
- 432Hz & 440Hz 145
- Frequencies & Vibrations Introduction 147
- Cymatics 149

Questions on Consciousness 157

Amissio Status Quo 163
- Thy Black Mirror 165
- The Illusions of Perceived Realities 167
- Sun 169
- Perception is Reality 171

Essence of the Ending Sublime Conclave: 173

Acknowledgements

I would like to thank my family that has so inspired me, since birth, to live in constant pensive thought and to always seek further light in the world.

I would like to thank the many master teachers and ancestors that laid the foundation for me to build a stable structure upon. Imhotep, the African, Dr. A.A. Yosef ben-Jochannan, Dr. John Henrik Clarke, Dr. Ivan Van Sertima, Dr. Cheikh Anta Diop, Dr. Frances Cress Welsing, Anthony Browder, Neil deGrasse Tyson, Bill Nye, Nikola Tesla, Albert Einstein, and so many more master teachers and ancestors that I have not mentioned.

I would like to finally thank the many other brothers and sisters like you that find great curiosity, pleasure, and purpose in the sublime arts. I wrote this book because we humans, as Bill Nye and Neil DeGrasse Tyson said need more inspired thinkers, more scientists, and more people that want to find the answers to the hardest questions known to humankind and are willing to partake in that beautiful journey in a pure heart. We need more researchers ready to spend the sleepless nights invested in chipping away at the wall of our ignorance.

Preface

Close your inhibitions and open your mind.

I want to inspire you to explore the sublime arts. I want to destroy your illusions. I want you to challenge everything and think critically. I want you to simply ask critical questions and to live in pensive thought. I want you to simply close your inhibitions and open your mind. Let the energy of the universe within you and learn from her majestic nature. We were not birthed here to pay bills and die. We were not birthed here to live in perpetual debt or struggle or powerlessness. We were not birthed here to succumb to the distorted distractions of the matrix; blinded by consumerism and searching for happiness within it. The answer to what we are here for lies in pensive thought. It dwells in contemplation, and it is there we must seek her.

Introduction

This book is twofold as it is a combination of the beautiful aspects of life that I have come across and my overall perception of the world that we all experience together; which I write from a mystic [*mystic, meaning to live in deep-thought and self-contemplation*] perspective.

I cannot recall a time that I was ever void of my thirst for the sublime arts. While I wasn't much aware of those sublime arts at the time, I was still curious as it related to the inner workings of it all - the universe and all of its majesty. I wasn't even knowledgeable as to what the sublime arts were to be entirely forthcoming. It wasn't until recently, about eight years from the first printing of this book, that I learned about the philosophy of Indigo children, the YouTuber known for his *Spirit Sciences* collection [*and I don't endorse his validity or insolvency*], and the movement of mass consciousness that has happened upon us all because of the interconnectivity of our computing devices. In this new wakefulness, I took a step back and realized that ever since we have had access to the internet and more information than at any point before, [*through mechanisms like Amazon's legendary logistics systems, or the ability for us humans to link massive amounts of data together electronically for sharing, categorization, and debate*] excluding the library of Alexandria, that we were destined to indulge ourselves in these sublime arts; binging them like hardcore Netflix subscribers watch entire television series seasons in just a few days. In placing myself more into the vast knowledge of the many realms of these sublime arts over the years, I have come across so much incredible information that has altered my perspectives, changed my actions, risen my thought process, and driven me mad with questions. The information I will be covering here in this book will be information that is liable to destroy your illusions of how you perceive that which is within us and around us, and that's the point. It is because of this reason that I bring you this book to share with you a few breadcrumbs of what I have found to be the inspirational concepts in the high sciences. It'll come in the form of nuggets that I refer to as the "Essences" and, as I stated earlier in this introduction they are twofold.

The introductions of the Essences at the beginning of each chapter are my thoughts, sensibilities, and perceptions based on reason, research, and contemplation. Within the actual subsections of the Essences, you will find that the information I present is more scientific, scholarly, and mostly without my subjective view. I've composed it this way because I want to show you how beautiful it is to remain in pensive thought and how sublime it is to study the nature of that which is effortlessly present; that is to say the world within us and around us [*I will use that phrasing so often that instead of always writing out "around us, within all" I will only write "auwa".*].

You will notice that I oftentimes will begin writing in a more poetic sense; ignoring the standard rules of grammar and such. I do this because I've learned that in writing or speaking in a more poetic sense the beauty of words and the ideas they help to convey becomes free from the formality that so governs the languages we speak. That being said, let us get to the nutrient-dense portion of the introduction which will give way to the first Essence, the Essence of Numerology.

There are so many incredibly beautiful aspects of life that we should be exploring. Why does the Sun perform the way that it does? Is the sky blue? Are what we call "colors" actually present? What is time? What is our sub-conscious mind? How does our body know how to do certain functions without our conscious control? Where does life originate? How immense is the universe? Why can't human beings seem to quit bickering over land masses and resources that no one can truly own?

The questions of why it was just recently that humankind had a manmade object reach the point of "interstellar space", or what exactly makes up the matter that makes up the matter are relevant to our reality and yet many of us aren't even aware that these are questions currently without an answer. There are so many more questions and whether they exist in the realm of history, science, mathematics, physics, or any other realm we should want to know what's happening auwa. Not just so that we can have some silly one up on another nation or people, or so that we may gloat about it, or so that we may aim to create some ultimate militaristically driven hell-on-Earth weapon capable of utterly destroying all forms of life on this planet – nay. And I'm not naïve to the possibility of evil actors acting out of ill will, irrationality, or contempt for another and I'm not so naïve to believe that all of the world's current complex problems will so easily be solved. However, this reality shouldn't thwart our reverence for the enigmatic and sublime arts. For these enigmatic and sublime arts create what I, from my reason and research deduce as the common good.

Our Creator has endowed us with a brain capable of doing far more than absorbing junk television, junk music, junk politics, and bathing it the negative energies of our food choices. We should wonder and aspire to know what's happening auwa because of the simple fact that when we know better we do better; you know, practicing dharma through innerstanding [*innerstand(ing) – this is a term that isn't in the dictionary but in this book it means that you correctly internalize, analyze, or process an idea or a concept. It sounds better than saying "understand" which, to me, assumes that I am underneath knowledge; nogo*]. We should aspire to learn more and wonder more because by existing on that thin line between our innerstanding and our ignorance we continually push back the lower natures of our thoughts, and our actions; again, practicing dharma through innerstanding.

{"When one is young, righteousness is instilled by fear. When one is in the middle, it's by choice. When one obtains the light, he performs Dharma through understanding."

– Abbot, *The Man with Iron Fist*}

This book isn't some fantastic utopian fairytale from my mind either as history is riddled with examples of what happens when knowledge in proper form guides our motivations.

You know, there is a statue in Kemet [*Kemet or KMT was the original name of the country now known today as Egypt. Kemet is translated to mean "the black land," "land of the black," or "land of the black soil"*] that many people know as the Sphinx. The original name of that statue was HOREMAKHET, which means Heru on the horizon or Heru of the horizon. It has the head of a divine man, and the subdued body of a lion and it was meant to serve as a reminder in stone that the divine mind of man should always subdue the beastly nature of man. It is important to note that "man" is a Sanskrit word that means "mind" but I will explain this further on in the book. The title "Sphinx" was placed on the statue by the invading Greeks that looked upon it and thought it was like the Sphinx from their legends. The word Sphinx means to strangle or to hold, and that definition derives from their story of Oedipus and the Riddle of the Sphinx. In this story, the Sphinx, an ugly winged woman beast, devoured all travelers who could not solve her riddle correctly. The Sphinx and HOREMAKHET have nothing in common aside from the fact that the Greeks stuck a label to a monument which they didn't innerstand.

By my curiosity and with the help of others of like mind we found this knowledge. Where would I be without it? Ignorant and unaware. I say all of that to say that if we allow ourselves to become complacent with the information presented to us without question, analysis, or research, we become part of that lower beastly nature of mind. It is only by proper categorization of history, and the study of the sciences that we as a species operate better in peaceful harmony with the biosphere and with each other. I, like many used to believe that the statue in Kemet was a Sphinx, and I didn't question it. I used to think that the name of Kemet was indeed Egypt, and I didn't question that either. I didn't know that Africans had a massive, peaceful, and prosperous civilization long before there was Europe, Greece, Rome, or any other nation. I used to believe that Christopher Columbus discovered America. I used to have no knowledge of the mysterious lore that is the ancient Lumerians or Atlanteans but in finding the sublime arts I have become aware of the beauty that is knowingness.

In a more modern example, before the 1970's the United States of America utilized leaded gasoline in to achieve higher octane ratings in combustion engines. As far as anyone knew at the time, this was an excellent way to achieve better fuel efficiency and engine power. The United States also utilized lead in household and toy paints because it sped up drying, maintained a freshly painted appearance, and resisted corrosion, which were all great benefits. However, it was through much scientific research, and curiosity that the world discovered something appalling. We found out that leaded gasoline excretes a pollutant that was directly responsible for central nervous system damage and impairing neurological development in the youth; a truth we could not ignore.

Imagine for a moment that no one ever investigated how deadly lead was, whether it be in the form of paint or in the form of fuel. How many children and subsequent adults would have grave disabilities in our current lifetime? A grim hypothetical that could have been true had we not been curious and driven by the genuine spirit of knowing. This is why a constant thirst for the curiosities of life is essential for the betterment of our immediate universe.

Another instance of humankind crushing its ignorance is with the birth and greatness of a man called Nikola Tesla. Before Tesla created alternating current electricity protocols, the world used Thomas Edison's direct current. Direct current based electricity, as we now know is highly inefficient. It was through Nikola Tesla's pure-hearted passion for science, technology, helping humanity, and breaking the spell of human ignorance that he was able to theorize that all energies are cyclic and thus utilizing an alternating current is more efficient and better suited to the natural nature of energies.

Imagine if the enigmas of life never inspired Nikola Tesla. Granted, he lived in a time where others who weren't as inspired as he constantly ridiculed him and left him as an outcast but imagine if those same people instead of ridiculing him empowered him and each other! What great creations might we have had today that instead we have been so denied? Sadly, we will never know and we are instead forced to speculate.

In a mystic focus, we are coming to innerstand that the Great [and I hate to use this word] Pyramid at Giza is in fact not a burial tomb at all and that it may have had some higher purpose such as a grand temple, Stargate, or cosmological observatory. To this very day, we can't even say how the ancient Africans even built such a magnificent structure let alone, what it meant to them. If human greed and discontent with each other for silly reasons was never allowed to prosper, we might have even still had the ancient Africans around to tell us firsthand what it all meant and how to read the Medu Neter. Reason and respect must guide us without the allowance for greed and corruption to infect our motivations.

It is our responsibility to our better nature to seek out the answers to the challenging questions of the ages, to record and remember those questions and answers, and to not be so easily pacified solely by the explanations of scripts written a millennia ago. History has shown us that by allowing curiosity and the thirst for knowledge to grow within us, that we find a better moral standing, and a better existence for ourselves, and for our environment.

In this book, I have compiled just a few aspects of life that we as a curious community have come across. I have also added a few notable quotations, poems, and mindsets to consider which are either peppered throughout the chapters or around the end of this book. It is my hope that they inspire you to learn about the many other aspects of life that I refer to as the inspirational concepts in the high sciences. I've written this book for all ages and all peoples in the hope that more will become inspired by these sublime arts and allow themselves to become removed from the simplistic desires of the matrix.

Essence of Numerology:

– How Numbers Guide Our Perspectives –

Numerology is the realm of knowledge that focuses on the significance of numbers and how those numbers relate to particular events in life. In short, Numerology is the poetic language of relationships. A numeral within the scope of Numerology is a life event or a circumstance within nature equated with a numeral. It serves as a teaching point, a charting point, or a mysticism-based point meant to explain something in the perceivable heavens.

Since the earliest of observable times, many of the ancient high science societies have studied and paid great reverence to the esoteric and sublime nature of numbers. This is seen from the relatively modern revelation that "7" is the number of completion to the fact that the numbers One through Seven, no matter what their ancient or modern names may be, have been utilized by countless ancient societies to assist them in living their lives. Whether to chart and measure their birth cycles and menstrual cycles (as the Lebombo and Ishango bones have shown us) or whether they were utilized to help with trade and barter, numbers have undoubtedly had a profound impact on all of our lives. I would submit that mostly all if not all of the ancient societies studied and paid reverence to the way that numbers shaped the world around them and within them. This reverence to numerology also shaping the perspectives that they internalized and which guided them in their daily endeavors. We all owe the ancient Africans a great deal of respect and adoration for their contributions in this sphere, for they delivered to the world the beginning; the light.

Numerology, in the modern perspective, is crucial to innernating how the societies of ancient times past were able to comprehend the most minuscule design features of the world around them and why it is something that we should continue. Numbers explained to them, and still do to those of us that are aware, the nature of the universe itself. The ancients created this entire system to assist them in everything from spiritual customs, planting customs, mystic customs, and many more aspects of life. They were able to tie these numerology concepts into the very core of their existence, and it is something that we still do today in order to simply explain and operate inside of the matrix we reside within. It actually reveals just how ingenious they were with innerstanding the divine nature of things. It is through this innerstanding of knowledge from the past that we are able to properly progress humanity further with intelligence, true heart, and in proper form. It is thence our responsibility to utilize the massive brain endowed upon us by our creator to seek that ancient knowledge, combine with it current advancements, comprehensively study this science, and teach it to the generations that come after us.

A great ancestor and master teacher Dr. John Henrik Clarke, (an honorary SAAMR, AMMATA & AMWA recipient) famously stated that "History is not everything, but it is a starting point. History is a clock that people use to tell their political and cultural time of day. It is a compass they use to find themselves on the map of human geography. It tells them where they are, but more importantly, what they must be."

− Dr. John Henrik Clarke

Numerology has inspired me to travel further on the path of innerstanding more about the relationship between numbers and the universe [*or potential multiverse*] around me and within me. It is because of my reverence and respect for the histories of many cultures of the past days that I refer to myself as a "Photonic Conduit" which means a "particle of light; gateway of light." I want Numerology to do the same for you, and so I have created this Essence of Numerology that should serve as an introduction into the mastery of the study of the significance of numbers both ancient and modern.

Authors Note for Essence of Numerology:

As you read this Essence, make certain to keep a notepad nearby, or at the very least a pencil or highlighter to notate sections of this Essence that are important to your overall innerstanding of why Numerology is such an important concept in the high sciences and why its principles should be deeply integrated into your daily lives. You also might notice that there is an Essence in this book called "Mathematics", and you may wonder why I separated Numerology from it – I want to explain this – what I did was separate the numbers themselves [*i.e. 1,2,3,4,5,6*] from the mathematic process [*i.e. 1+2+3+4+5+6*]. What I did is isolated formulas, equations, and processes within mathematics and gave that its own section for reasons that will be revealed to you when you reach that Essence. Numerology, however, I cannot stress enough, is more about the (esoteric) significance of numbers themselves and not the equations, formulas, and other mathematic processes associated with them.

One last note before you begin this Essence; the Hindu-Arabic Numerals are the numbers 1, 2, 3, 4, 5, 6, 7, 8, and 9. The number "0" was added some time later and thus you have what we in the modern era utilize as our numerals system. The Hindu-Arabic numerals were created by the Indian mathematicians and the Arabic mathematicians of times past who created this system by way of their own creative intelligence and the knowledge taken from the ancient African societies that predated them. In the process of analyzing Numerology I will be utilizing this Hindu-Arabic system, however, do bear in mind that other societies had their own way of representing the numerals that I will not explore in this book. These other systems are worth finding out; just on your own time – I'm just here to plant the seed in the garden, water it, and make certain the Sun shines on you; it's on you to grow. Turn the page to begin your wonderful adventure with me into the realm of Numerology and beyond.

{"Open your minds. Perspectives change when knowledge is obtained"

– Cedric Paul Harriott (Heh Heru)}

From the Everything of Nothing – 0

From the everything of nothing is a phrase I created that is the embodiment of how you should see the numeral Zero. Zero is an oval, in our current application and that esoterically holds the void that is nothingness and yet the wholeness of everything within its symbol. Think about it for a second and let the concept of what I am saying change your perceptions of this numeral. The numeral "0" quite literally represents the idea that from the so-called great void of nothing comes everything. Zero represents the "Cosmic Egg", the placeholder value, or the vacuum that is the perceived emptiness of space, which holds the potential and ingredients for all of life to blossom fruitfully. Zero highlights the void of life, and yet the entirety of life all beautifully contained in its oval. I must continue to reiterate how important it is to visualize this concept: Zero represents the cosmic egg that contains all of whatever comprises the sub-atomic particles that allow for our current form of life to exist. Zero is the point in time where we cannot perceive life, and yet it contains all of the ingredients of life.

You can even see how it encompasses this concept by looking at the Fibonacci numbers series: 0, 1, 1, 2, 3, 5, 8, etc. If you aren't familiar with the Fibonacci series, don't fret or Google it as I will give you a brief synopsis of its concept in this Essence and go much further in detail within the Essence of Mathematics.

The name of the Fibonacci sequence is named after the merchant and traveler Leonardo of Pisa that brought it to Europe in the early 13th century. He didn't actually discover it as it was used in ancient India where he stumbled upon it. It is important to note that before it was found in ancient India, it was also in ancient Africa; a civilization that predates the ancient Indians by thousands of years. I will, however, use the name Fibonacci sequence out of ease of research value.

The Fibonacci sequence is a mathematical equation that creates a special spiral at a special degree that many (if not all) things in nature follow. Hurricanes, the Milky Way Galaxy, Pinecones, Seashells, the Human Face, and many other examples all follow the Fibonacci sequence and are represented utilizing numbers or a graph. In all of these examples of the Fibonacci sequence in nature they all start from the mathematical Zero and spiral outward into their various forms of matter.

Consider this, before you were a born child, before you were a fetus in your mother's womb, or before you were a zygote in your mother's womb, or a sperm in your father's scrotum, or the matter that makes up what a sperm is, or the matter before even that, or even before that matter that made up the other matter, you were a zero - a nothing and yet you were the possibility of life – because some divine things happened to bring you into existence.

Another example is how before your car can go One mile per hour you must come from the nothingness of speed that is Zero.

Or before you can master a language you must start at level Zero, or no innerstanding at all, i.e. nothingness.

{"When one's expectations are reduced to zero, one really appreciates everything one does have." – Stephen Hawking}

And Then There Was One: 1

Many ancient societies represented the beginning of human experience as the mathematical number One, but I will not be spending that much time in exploring those societies as the Internet and libraries are plentiful in their exploration of those societies. However, I will instead inspire you to see the number One in a completely different way. That is to say, the way that these ancient societies perceived it. I get my innerstanding of the number One from an Ancient African perspective (also called a "Pythagorean Perspective") [*Pythagoras who received his perspective from his own wisdom and the wisdom of the ancient Kemites of North Africa that predated him.*] and in that perspective; I view the number One as God. God to me isn't a person, or even a spirit but rather a title that signifies the Creator or the Grand Architect of the Universe; Creator also, not necessarily meaning person or spirit but rather whatever has brought all that we view into existence. I see the number One as simply signifying that from which all things come. Bringing this full circle you have the Zero which is the void or the Cosmic Egg, and from that Void of nothingness, you get that from which all things come from; the number One – Thy Creator.

Change your perspective on this number. Realize that in all things around you there is more – there is a deeper meaning. You can view this number as simply a means to an end; something that helps you accomplish finding out whether or not the cashier at the grocery store has given you the correct change OR you can view this number from a more esoteric sense like the ancients did and see that it is synonymous with the Creator, with the beginning, and with life. The colors you begin to see when your perspective shifts into the latter become far more vivid and beautiful. This book will push your mind – let it. Open your mind.

{"It's harder than sitting with a blind man and trying to describe yellow."

– Lupe Fiasco, *Theme Music to a Drive By.*}

The Masculine & the Feminine; Duality: 2

The number two in ancient societies represented everything that has an opposite or counterpart. The number two represents the equality of dual balancing forces. For example: Yin & Yang, Sun and Moon, Heaven and Hell, Man and Woman, Good and Evil, and so on. One of the most noticeable instances that this can be seen in is with the Taijitu; the symbol for Yin & Yang. This concept in Numerology has not changed and means inherently the same today. In my own logic (and through my own mental court of reason) you cannot really have Good without Evil, or Light without Darkness; they are in a synchronous dance that highlights the primary concept in the number two. That concept is BALANCE. The number two is symbolic of how everything should be in a natural balance in perpetuity.

For example, this is a concept that is seen in the ancient Kemetic story of Ausar and Auset. They are man and woman and from that unity of man and woman you receive the number 3; child – meaning life in perpetuity essentially. For the ancient Kemites this is a way that they taught their culture the concept of dual balance; the masculine and the feminine. It played such an important role to their development that they integrated it into one of their most sacred stories.

Another example is from the ancient (and even modern Chinese peoples) where "Yin Yang" describes how contrary forces are actually in a balance or complimentary towards each other's survival. It portrays their interconnectedness to each other and was so important to the Chinese way of life that it was integrated into the very core of their philosophical thought and in traditional Chinese medicines.

{"You must be shapeless, formless, like water. When you pour water in a cup, it becomes the cup. When you pour water in a bottle, it becomes the bottle. When you pour water in a teapot, it becomes the teapot. Water can drip and it can crash. Become like water my friend."

– Bruce Lee}

Bruce Lee innerstood the concept of this duality through the numeral 2 and employed it into how he mastered his art. He employed the idea of balance throughout his martial arts career by internalizing the concepts of perpetual balance; that energy is cyclic. If you watch his martial arts displays through his videos you can pick up his tuning into this ancient knowledge and his application of this ancient knowledge. You can watch how the Shaolin monks perform their daily exercises and teachings which revolve around the same core concept of being in due balance with all mankind. There was a television show called *Avatar: The Last Airbender* where the entire premise of the show was to keep the worlds many inhabitants in balance with each other and the spirit world. I advise you to watch the Shaolin monks, and watch that television series; tuning in on this concept will become much easier with the help of that visual aid.

Trinities, Triangles, Third Eyes and the Child: 3

The child is signified by the number three in Kemetic mythology. From what I gather of it, the unity of Ausar (the father or phallus) and Auset (the mother or womb) produce the perfection or the Child; Heru. This concept is also in the 3-4-5- Triangle where Ausar can be represented by the upright side (or the length), Auset can be represented by the plane side (or the width), and Heru (their child) can be represented by the hypotenuse (or the slanted side). Bringing it full circle again Zero is the void and the egg, one is the creator, and two is the unity of the mother and father, which brings in the child. Many societies innerstood the biological fact that from mom and dad you get either boy or girl and from that the boy or girl they would find another boy or girl that would create yet another boy or girl and the cycle of creation would thence continue. Mother + Father + Child is a trinity of life represented by the number 3; this concept was an essential part of how this ancient society saw the immortality of life through the co-builder Gods (mother and father) forming a balance. This shaped the entire way that they functioned and was so profound that they integrated this concept into one of their most revered stories of creation.

What do we usually refer to the Pineal Gland as? The Third Eye is the answer. This is a concept seen in African cultures (The Eye of Heru is an example), Indian cultures (The Chakras are another example), and in many other cultures as well. It is the concept that points to seeing with the most sacred portion of the Mind; the Pineal Gland – the gland believed to be our key to the metaphysical universe. This is also the gland that science is still discovering new things about every day but that somehow many ancient societies knew about for thousands of years and revolved their lives around.

{"We have five senses in which we glory and which we recognize and celebrate, senses that constitute the sensible world for us. But there are other senses – secret senses, sixth senses, if you will – equally vital, but unrecognized, and unlauded."

– Oliver Sacks}

Back to the trinity concept of mother + father = child, where are they to live is the next question. They must have a foundation to build their home and lives upon – and this is where the concept of the Masonic square comes in as represented by the number four.

The Operative Builder; the Masonic Square: 4

The square, being the foundation that we build our lives upon, this is where we humans live – our physical reality - our physical plane to which we are bound. And it is in this awareness that we realize the Earth that our feet walk upon is our home. A square having four corners (and being a flat figure) is why the square is represented by the number four in both ancient numerology as well as in modern numerology – It is why one would travel to the "four corners of the earth" for whatever ends they wish to achieve.

Earth, we know from science and our ancestors is an imperfect sphere. However, the square represents this area in an esoteric sense that alludes to the flat observational plane to which we are grounded upon. This flat square enables us to perceive the other dimensions. Staying with our current three-dimensional perspective, this foundation allows us to perceive height, width, and length.

So as a quick recap:

- Zero is the nothing and cosmic egg; the boundary.
- One is the creator; whatever forces brought life to existence.
- Two is the masculine and the feminine; the mother and father that produce life in perpetuity that we observe.
- Three is the life that the mother and father bring in; it represents the child as either a male or a female.
- Four is the plane or the foundation; this is where we live – i.e. the Earth, our planet, or our groundings.

The interrelationship between these numerals have been designed, studied, perfected, reviewed, perfected again, and implemented by many societies as early as approximately 5,000 years ago [*possibly far more than that conservative estimate*] – and has endured a span lasting unto the current 21st century of thought we currently reside in. These are essentially shared human customs that have stood the test of time because of their ability to better guide and upgrade the consciousness of whatever society implements them within their lives. That's insight.

{"Wisdom dwells with contemplation: there we must seek her"

– Masonic Saying}

When the phallus and womb meet; life is birthed; 5.

Mind & the Womb: 5

The number five is a symbolic representation of man. The word "Man" is a Sanskrit word that means, "Mind" and the word "Woman" means "The womb that produces the man (or mind)." Five is also representative of the five senses that we humans in our current perspective know about, and it symbolizes the five fingers, the five toes, and the five parts of the body (the head, two arms, and two legs). According to Pythagoras, the Greek mathematician, the number 5 was the number that represented the perfect human microcosm. He was most likely alluding to the five fingers, five toes, five parts of the human body and even the five senses. Aristotle, a Greek philosopher and scientist, added a 5th element to the alchemical base (that alchemical base being Earth, Fire, Water, and Air). That 5th element was called "Aether" and was alluding to a more spiritual and metaphysical aspect. He was specifically referring to things such as stars in the heavens that must be made from something outside of the classical four alchemical base elements. It is important to note that both Aristotle and Pythagoras learned their craft from Kemetic teachings.

The number five was integral in many varying types of philosophical thought whether from the ancient Kemites of North Africa or the Greeks this number was held in a great regard. It guided much of their innerstandings, many of which we still analyze and utilize in our innerstandings today such as the five senses, the five parts of man, etc.

Numbers, as you can see now, have always had an esoteric'ness to them – an esoteric'ness that was considered to be a profound innerstanding by all ancient societies that grew to observe their benefit. It made life better, it made heavy concepts easy to grasp, it made concepts easier to chronologically order and also to more effectively teach utilizing the simplicity of numerals. It was and still is a sublime beauty worthy of our study – we must be good masters of this acquired knowledge.

{"Only the existence of a field of force can account for the motions of the bodies as observed, and its assumption dispenses with space curvature. All literature on this subject is futile and destined to oblivion. So are all attempts to explain the workings of the universe without recognizing the existence of the ether and the indispensable function it plays in the phenomena. My second discovery was of a physical truth of the greatest importance. As I have searched the entire scientific records in more than a half dozen languages for a long time without finding the least anticipation, I consider myself the original discoverer of this truth, which can be expressed by the statement: There is no energy in matter other than that received from the environment."

– Nikola Tesla}

Carbon + Death? + Life + Transcendence: 6

Carbon is a chemical element that humans are made of, and that also is prevalent in the atomic makeup of many organic things in nature.

Carbon has the esoteric atomic composition that consists of six protons, six neutrons, and six electrons; the true and beautiful meaning behind the 666 numeral grouping. While this number group has received an unfair negative reputation, the actual ways that this number is shown in nature is far more sublime and beautiful.

You know since the beginning of recorded times it has been noticed that many high science cultures (including our very own 21st century one) have buried their dead in a coffin, a six-sided figure that is carried by the usual six pallbearers. People also bury their dead under six feet of earth, and number six is also used to signify the death of the mind, body, and spirits experience in that earthly boundary (body).

Carbon, in large part grants us life, it leaves our earthly bodies after we die [*die as in physical experience but energy doesn't die. It simply becomes transferred into something else*], and becomes a part of some new life form, which alludes to our transcendence from one life form experience to another.

The concept in the previous paragraph paints a beautiful picture that serves to showcase the serious interconnectedness of all living beings – we are all made of star stuff. This number was (and still is utilized) to bring one's mind into a higher consciousness – or a greater level of thought. It allows one to realize just how interconnected all life in this universe is at an atomic level, and also it helps to highlight the fact that life and energy never die – rather they simply become a part of something else.

When I die, I am buried, and in time my body will succumb to the bacteria and other organisms that will break down the matter that has made me and utilize that matter in the survival and progression of their own lives – in essence, I become of them. The circle has no beginning, or end and this is the concept of Life, then Death, then Transcendence to Life, then Death, then Transcendence to Life and so on for infinity.

Something that I still search for is the answer to the question, where does life originate? What tells the atomic structures that make up the universe how to function? Who is Amen? [*Amen is the Kemetic deity that stands for "the Hidden One". He alludes to whatever life force animates our bodies*] It is safe to extrapolate given all that we know about the universe that there are forms of life out there and in that we find ourselves in another conundrum, where did it all come from? Such a massive question deserves more focus than it is currently receiving but I digress. Let us continue with the numerals.

{"For life and death are one, even as the river and the sea are one."

– Khalil Gibran}

{"Life and death are one thread, the same line viewed from different sides."

– Lao Tzu}

Completion & the Seventh Heaven: 7

The number seven represents Heaven. Heaven has had many different names, understandings, and interpretations throughout the years but one of the oldest concepts that I have found has come from North Africa's country called Kemet. In the Kemetic story after Ausar [*not the Greek name Osiris*] died he was resurrected and ascended into the seventh heaven to sit upon the throne of judgment. This was the first story of resurrection and predated all of the other subsequent societies that had their own iteration of this concept. However, we know that Kemet was the culmination of the wisdom and intellectual prowess from the whole of Africa towards the beginning of the Nile River, so it's safe to wonder if this system comes from some other place. I am still studying and researching to this day but as of this moment I haven't found many references to the concept of the 7th heaven outside of Kemet; I'm enjoying the journey though.

For the Christians, God rested on the seventh day and thus seven is seen as the number of completion for them. In other societies, their reverence and divinity for this numeral came from their own myths and legends that they utilized to help them break down heavy creation concepts.

The Voided: 8

The numeral 8 in numerology represents infinity beyond the 7th heaven.

Oh: 9

You'll just have to trust me on this one but the number 9 is far too sublime to get into right now. You'll have to wait until later on in the book before I reveal this light to you!

Kemetic (Egyptian) Numerology

The ancient Kemites integrated numerology into not only their mythological stories but also into the design of their structures in Kemet. One of the most famous of all of their structures is what people refer to as "The Great Pyramid at Giza". However, before I continue with this Essence in Numerology, it is important that you innerstand what is actually at Giza and why the word "Pyramid" is incorrect and will not be used in this book. After researching the etymology of the word "Pyramid," I have been able to prove that the word "Pyramid" is a misrepresentation of what is actually at Giza. Through my research, I have found that the word "Pyramid" is a Greek/Hebrew translation of what the ancient Kemites called their structure. Pyramid coming from the word Pyramidos and it alludes to the tiny pointy-topped cakes that the invading Greeks were reminded of upon looking at the megalithic structures built by the Kemites. I have found evidence to support the claim that the ancient Kemites called it MR (Mer or Mir) and it means "Place of Ascension," While I will not delve into the whole origin of the word you can follow my breadcrumbs by searching for the original Kemetic meanings of words as extracted from the source material left behind in Kemet (i.e. the tablets, papyrus, Rosetta Stone, etc.). From this point on in this book, I will be referring to what is at Giza by calling it "The Great Temple at Giza"; instead of "The Great Pyramid at Giza."

Let's examine some numerology concepts – The Great Temple at Giza (as a whole structure) can be represented by the number "One", which signifies the whole structure. The triangular faces of the temple can be represented by the number 3, as the temples triangular faces have three corners (angles). The square foundation has four corners (angles) and can be represented by the number four. And finally, the four corners of the foundation plus its apex have a total of five points. This is called "The Egyptian Triangle" or "The Egyptian 3-4-5 Triangle" and it is a foundation for geometry and many other high mathematic principles. The ancient Kemites had knowingly developed the concepts for triangles, squares, volume, points, angles, and so much more simply innerstanding that numbers have a great significance in explaining the world auwa. It is to these ancient scientists and the many cultures that learned from them that we owe a great thanks for helping develop the world we innerstand today. It is important to innerstand that because of the greed and war driven corruption of some cultures that Kemet fell and that its inhabitants were decimated. We must learn what happened here to accurately remember these cultures and to not ever repeat the evils that were done in the past to these great civilizations.

Let Us Break Bread: 0-9

Again, in the study of Numerology, we must remember that we are looking only at the individual numbers and their relationship to understanding concepts in life. While they do have a relationship with mathematic processes, they share more of a relationship with events in nature that are represented by the numerals themselves.

And so it was written that in the beginning came the creator consciousness, and from that consciousness, the creator created man and woman. From that man and woman comes the perfection or the child. It is from that child that the mother and father create the foundation for their family. That family, acting in unity creates a collective mind aware of the inevitable transcendence. From that transcendence, one is able to perceive the heavens separate from the physical realm. From that spiritual plane, one can perceive the relativity of time and is allowed access to the higher. From that relativity of time, one realizes the dimensions and the higher aspects of life beyond the bound physical dimension.

I want to explain it again with the numerals engrained.

And so it was written that in the beginning (0), came the creator consciousness, and from that consciousness the creator (1), created man and woman. From that man and woman (2) comes the perfection or the child (3). It is from that child that the mother and father create the foundation for their family (4). That family, acting in unity creates a collective mind aware of the inevitable transcendence (5). From that transcendence, one is able to perceive the heavens separate from the physical realm (6). From that spiritual plane, one can perceive the relativity of time and is allowed access to the higher (7+8). From that relativity of time, one realizes the dimensions and the higher aspects of life beyond the physical dimension and thence realizes that we are nothing more than the universe expressing itself in human form (9).

This explanation of the human condition from a metaphysical perspective is something that has been seen in every culture that I have researched. In some way or another, they all seem to have fragments of this same perspective, and it guided the way that they lived their lives. Where did it originate? The Kemetic natives of North Africa kept the mysteries intact no doubt and created great inventions in their dynasties but it had to have originated from somewhere earlier as Kemet was the culmination of the wisdom of the whole of Africa and this story I have found far before that. Did it come from the Dogon? The Ethiopians? The Nubians? The mysterious Lumerians or Atlanteans? I am still searching for the answers myself but we should all search for the answers, and enjoy the journey.

Our role in this is to internalize this concept and to aspire to attain access to the higher dimensions. Our role is to innerstand what the creator is, what the cosmic egg is, and how to erect ourselves to attain this knowledge in proper form. Our role is to not become distracted by the matrix of man but rather to be open and receptive to that which is truly sublime in the heavens.

Essence of Mathematics:

– How Mathematics Hold Such Grand Secrets –

Mathematics, according to Merriam-Webster is "the science of numbers, quantities, and shapes and the relations between them"[1]. I use only that portion of their definition (and suggest that you do the same) as it very clearly states what mathematics is in our current innerstanding. I also add that mathematics, in a more esoteric sense, is a language of the Creator – utilized in the design and functions of the universe. Mathematics has been around much longer than 1573 and is showcased in the ancient societies of the Kemites, the Mayans, the Olmecs, and many more societies that predate the current 1573 origins of mathematics date. I attended public school and only remember learning what I now call "Junk-Food Mathematics." I refer to it as such because I remember being taught math with the aim of passing state exams and not learning math to better appreciate the divinity of the whole of mathematics; this is what I will be covering in this Essence. Mathematics is everywhere and integrated into everything that we do and are. The Golden Ratio, Phi, Fibonacci sequence numbers, and the Great Temple at Giza, are examples of high mathematics in an ancient sense. In a modern sense, we can see how mathematics is so deeply integrated into the other sciences such as languages, chemistry, biology, and within the teachings of music. Mathematics is a language of code that beautifully describes how the things we observe work. That's insight.

---

[1] "Mathematics." Merriam-Webster. Accessed January 6, 2015. http://www.merriam-webster.com/dictionary/mathematics.

Pre-Fibonacci & Thee Fibonacci Sequence

In the year 1202 C.E. (meaning Current Era) the Europeans were introduced to a new branch of high mathematics by a mathematician and merchant named Leonardo da Pisa. Leonardo wrote a book titled, "Liber Abaci" meaning (The Book of Computation) and in this book he described the math systems that he had learned from the Arabic and Indian Mathematicians. He learned these new math systems from the Hindu-Arabic numeral system [The Hindu-Arabic numerals were the numbers 1-9]. It is important to note again that Leonardo did not discover or invent this system; rather, he acquired the knowledge from the Hindu-Arabic numeral system [Arabic & Indian Mathematicians] and the Kemetic [Egyptian/North African Mathematicians] additive series of profound dimensions.[2] Dan S. Ward in his article on the Fibonacci Numbers pays the proper respect to the line of great mathematicians that are usually neglected when speaking on the Fibonacci Numbers, he states, "Clearly our society owes a great debt of gratitude to Fibonacci – as well as the Arab scholars who kept the knowledge alive, and the Egyptians for holding the mysteries intact."[3]

Merriam-Webster defines the Fibonacci Numbers as "an integer in the infinite sequence 1, 1, 2, 3, 5, 8, 13, ... of which the first two terms are 1 and 1 and each succeeding term is the sum of the two immediately preceding [numbers]"[4]. This definition essentially describes the process of adding numbers in this format: 0+1=1, 1+1=2, 1+2=3, 2+3=5, 3+5=8, 5+8=13, and so on for infinity. As you can see what we are doing is adding the first two numbers in an

---

[2] Ward, Dan. "Fibonacci Numbers." Fibonacci Numbers. June 6, 2005. Accessed January 8, 2015. http://www.halexandria.org/dward093.htm.

[3] Ward, Dan. "Fibonacci Numbers." Fibonacci Numbers. June 6, 2005. Accessed January 8, 2015. http://www.halexandria.org/dward093.htm.

[4] "Fibonacci Number." Merriam-Webster. Accessed January 8, 2015. http://www.merriam-webster.com/dictionary/fibonacci number.

additive and sequential progression (ex. 0+1=1) then, for the second step we are adding the number immediately to the left of the sum (ex. If 0+1=1, then to find the next Fn [Fn = Fibonacci number] we must add 1+1 which gives us 2. Using this method, in order to find the third Fn we must add 1+2, which would give us 3. Using this same formula, in order to find the fourth Fn we must add 2+3, which would give us 5. We would then perform this same process over and over which provides us with the next Fn and this is how you compute the Fibonacci Series Numbers.)

These seemingly meaningless numbers would indeed be meaningless had nature not been using this same process to create things like Pinecone spirals, Pineapple spirals, Seahorse tails, Hurricanes, Sunflower Seedpods, Spiral Aloe Plants, the Nautilus Shell, the Milky Way Galaxy, and so much more. To keep your eyes in this book, I am including a few images that show you this beautiful showcase of high mathematics sprinkled in so many forms within nature.

"Aloe Polyphylla Schönland Ex Pillans." Flickr. August 16, 2006. Accessed January 8, 2015.

Eberle, Scott. "Re:Play Blog." RePlay Blog RSS. January 19, 2012. Accessed January 8, 2015.

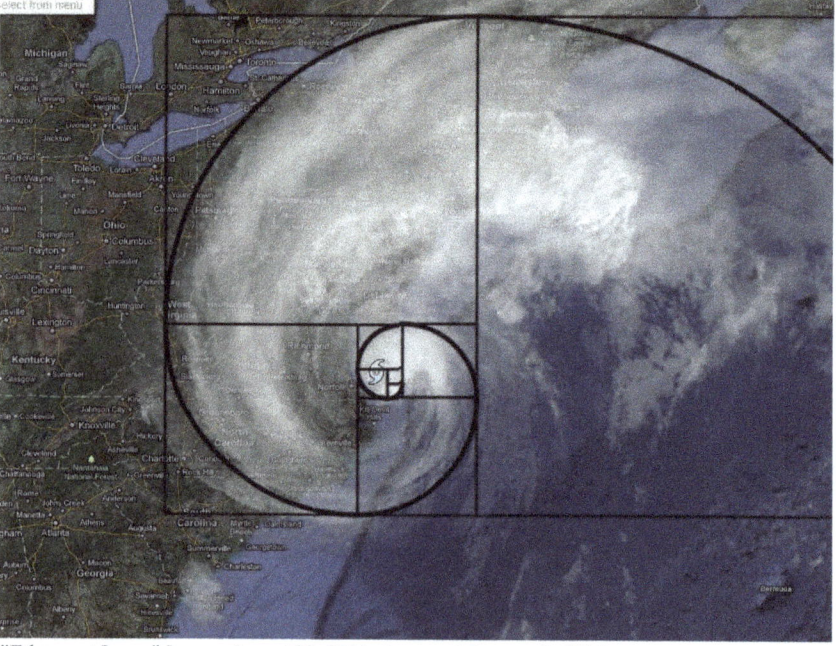

"Fibonacci Irene." Imgur. August 28, 2011. Accessed January 8, 2015.

The Fibonacci sequence is an inspirational concept in the high sciences due to its relationship with the process that nature uses to create many things around us as well as its ability to be seen using numbers and mathematical processes. It is as if we have a key into the innerstanding of what the creator used as a design feature in the development of the cosmos and all of its living species. How sublime is it to see that relationship so clearly? Doesn't it just inspire you to want to explore more about where this process came from, why nature is using this formula? Who or what created the processes that nature uses to develop the many interconnected areas of life? It is an inspirational concept because we have discovered so much about this and yet it shows us researchers and scientists how very little we innerstand about that which is happening within us and around us; THAT my dear friends is insight.

The Fibonacci sequence also has a twin brother who's in the form of a unique number. That twin brother's name is the Golden Ratio but they also answer to the Golden Mean, Divine Proportion, the Golden Section, or the Golden Rule; all of which mean this same number: 1.618.

The Golden Rule / The Golden Ratio: 1.618

The Golden Ratio is a mathematical process that gives you a particular number that is found in many of the same areas in nature as the Fibonacci sequence. The reason for this similarity is because the Fibonacci numbers and the Golden Ratio give you the same design outcome found in nature's many living organisms. In order to calculate this special number one must follow a precise mathematical process, one that has been repeated by the ancient Kemites of North Africa, the Greeks, and many other cultures around these two civilizations and after them in the chronological order of recorded civilization.

The equation for the Golden Ratio is:

$a/b = (a+b)/a = 1.6180339887498948...$

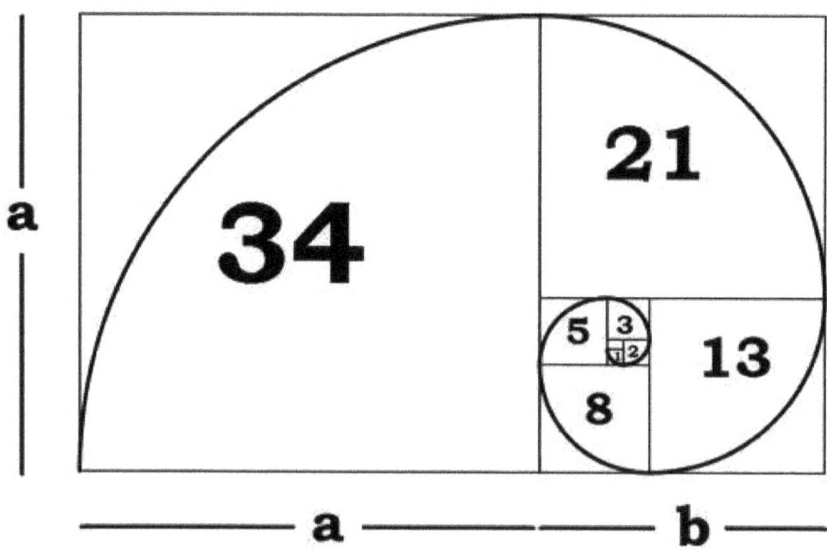

J. Elaine. "What is the Golden Ratio?" LiveScience. June 24, 2013. Accessed January 10, 2015

According to LiveScience's Elaine J. Hom, "The Golden ratio is a special number found by dividing a line into two parts so that the longer part is divided by the smaller part is also equal to the whole length divided by the longer part."[5]

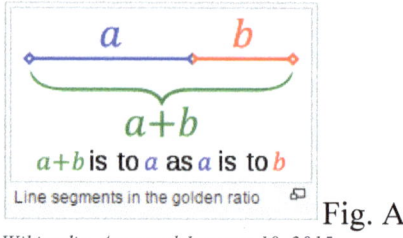
Fig. A
Wikipedia. Accessed January 10, 2015

Fig. B
Wikipedia. Accessed January 10, 2015

The number 1.618 is mathematically shown as the Greek symbol for Phi (ϕ). What many of the other ancient societies called it or how they symbolized it is knowledge that is not that clear at this present book publishing but as with all things in the realm of knowledge nothing is lost forever; at least that's what I'd like to believe.

---

[5] J, Elaine. "What is the Golden Ratio?" LiveScience. June 24, 2013. Accessed January 10, 2015. http://www.livescience.com/37704-phi-golden-ratio.html.

One of the oldest monuments that still has much of its properties intact and also the Golden Ratio within its contents is the Great Temple at Giza. At its current measurement, the length of each side of its base is approximately 755.75-756 feet (or 230 meters) along with a height of approximately 481-481.4 feet (or 147 meters). From its base to its peak this gives us an approximate ratio of 1.5717.[6,7,8] -- I, from my research, have gathered sufficient consensus that since this is so close to that Golden Ratio that the ancient Kemites of North Africa must've known about and utilized the Golden Ratio in their daily lives. We have to realize that this Temple at Giza is around 4,600 years old and in that time many invaders have come into the land of Kemet and chipped away the Great Temple's limestone shell for their own monuments, drilled holes into it looking for treasure, and all sorts of other horrible acts of disrespect. We must then assume that through these many periods the dimensions became slightly offset from not being maintained by North Africa's original monument builders in that land.

The Golden ratio was also thought to have been utilized by the Greek sculptor and mathematician named Phidias (500 B.C. – 432 B.C.) in his design for the Greek Parthenon. Another Greek high science connoisseur by the name of Plato (428 B.C. – 347 B.C.) paid such great reverence to this golden number that he considered it to be one of the most extraordinary and sublime concepts in all of the mathematic relationships. Euclid (365 B.C. – 300 B.C.), another Greek native alluded to the existence of Phi within the pentagram (as shown in Figure C).

---

[6] J, Elaine. "What is the Golden Ratio?" LiveScience. June 24, 2013. Accessed January 10, 2015. http://www.livescience.com/37704-phi-golden-ratio.html.

[7] "Pyramids of Giza | Pyramids, Egypt." Encyclopedia Britannica Online. Accessed January 10, 2015. http://www.britannica.com/EBchecked/topic/234470/Pyramids-of-Giza.

[8] DeSalvo, John. "Measurements of the Great Pyramid." Measurements of the Great Pyramid. Accessed January 10, 2015. http://www.gizapyramid.com/measurements.htm.

Figure C. $\quad \dfrac{\text{red}}{\text{green}} = \dfrac{\text{green}}{\text{blue}} = \dfrac{\text{blue}}{\text{magenta}} = \varphi.$

Alchemistra. "Sacred Geometry-The Pentagram." – Esoteric Online. June 30, 2009. Accessed January 10, 2015.

Elaine J. Hom, a contributor from LiveScience has found some incredibly sublime areas where the Fibonacci sequence and the Golden Ratio reveal themselves. Her ten beautiful examples are:

- Flower Petals: The number of petals on some flowers follows the Fibonacci sequence and Golden Ratio.
- Seed Heads: The seeds of a flower are often produced at the center and migrate outward to fill the space. As an example, sunflowers follow this pattern.
- Pinecones & Pineapple Scales: The spiral pattern of the seed pods spiral upward in opposite directions of the Pinecones and Pineapple scales essentially follow this same design.
- Tree Branches: The way that tree branches grow is an example of the Fibonacci sequence and Golden Ratio.
- Shells: Many shells, including the snail shell and the nautilus shell, are perfect displays of the Golden Ratio and Fibonacci sequence in nature.
- Spiral Galaxies: The Milky Way has a number of spiral arms everywhere, each of which has a logarithmic spiral of roughly 12 degrees. The shape of the spiral is identical to the Golden Spiral, and the Golden rectangle can be drawn over any spiral galaxy.
- Hurricanes: Hurricanes oftentimes will display the Golden Spiral and it is usually visible by looking at the space images of the cloud formations they create on earth.
- Fingers: The length of our fingers, each section from the tip of the base to the wrist is larger than the preceding one by roughly the ratio of Phi.
- Animal Bodies: The measurement of the human navel to the floor and the top of the head to the navel is the Golden Ratio. But we are not the only examples of the Golden Ratio in the animal kingdom; dolphins, starfish, sand dollars, sea urchins, ants, and honeybees also exhibit the proportion.
- DNA Molecules: A DNA molecule measures 34 angstroms by 21 angstroms at each full cycle of the double helix spiral. In the Fibonacci series, 34 and 21 are successive numbers.

Source: J, Elaine. "What is the Golden Ratio?" LiveScience. June 24, 2013. Accessed January 10, 2015. http://www.livescience.com/37704-phi-golden-ratio.html.

The incredibly sublime Golden Ratio and its twin, the Fibonacci Sequence are an inspirational concept in the high sciences because they exist in the very molecular blueprint design of so many interconnected things within our eyesight on earth, and within that which we are discovering more about with every passing second; the cosmos, the heavens, the metaphysical – that which is beyond the realm of physical.

Why should this be important to you? The answer to this question is in the form of a saying that I came up with when I became more aware of that which is happening around us and within us and that phrase is "Perspectives change when knowledge is obtained". How you see yourself, how you see me, how you see the spider, the rat, the rocks, the dirt, the water, the sky, the moon, the colors, the lights – how you experience the wind, the meditation, the feeling of heat, the sounds of silence, and so much more are completely changed when you innerstand what is happening around us and within us and how much we know about how little we know!

For me, when I became more aware of that which is AUWA I found myself staring more at the sky at night, breathing in the cool breezes and looking at the constellations, the spots on the moon, and the bright flickering lights in the night sky. I have noticed that as I learned more about the moon's powers of pulling tides I would ask myself that if we humans are made of so much liquid material and the moon can pull water I wonder if it pulls on us, I would wonder just how many galaxies are actually within space and how little of them that we can actually see. I would oftentimes ask myself why haven't other forms of life reached out to us? Or have they? I would ask myself what is really at "Challenger Deep" [*the deepest part of the ocean*]; I wonder why we humans haven't figured out a way to truly go that deep into the ocean or why we aren't paying more attention to how Voyager 1 and Voyager 2 are reaching parts of space that we have no understanding of [*Voyager 1, launched September 5th, 1977 just recently reached interstellar space; an area of space that we have very little knowledge of thus far*]. This is what your mind does when you let in these inspirational concepts in the high sciences. They open your mind and dare I say that help you open your Third Eye and Crown Chakra – it is a euphoric rush of inspiration for how you live your life and in what direction you want to take your life.

It somehow makes taking selfies, listening to negative music, or watching junk television programming less fulfilling.

I was watching Cosmos: A Spacetime Odyssey hosted by Neil DeGrasse Tyson and in that he mentioned that Carl Sagan invited him to experience some high science (the Cosmos). Neil, who was then just a child from New York knew from that moment that he wanted to be in the field of Astrology and if you fast forward from that moment all of these years later he is known as one of the most recognized names in that field next to Bill Nye [Bill Nye was my Carl Sagan to Neil – After my parents caught wind that, from a young age I was always into learning something sublime, they bought me the Bill Nye the Science Guy Computer Disk game that I played religiously until it broke. Needless to say, that game all of these years later I can still recall the music and everything for]. I hope that from this book of inspirational concepts in the high science you go further and research these deep topics that I am covering; be inspired to always reside in the search for those high sciences – this is how we grow and as I always say at the end of a powerful concept; "That's Insight".

The 9 Code; Its Sublime Insertion into Angles

Told you I'd come back to you about the number 9!

The 9 Code is the name of a mathematical process that alludes to how the number 9, the last number of the Hindu-Arabic numeral sequence is found in the reduction of degrees of an angle or degree of a circle. The mathematic process of reducing a number is first adding the numbers of a degree or an angle to find the digital sum of that number and you continue adding these numbers until you are reduced down to a single digit, which in the case of the circle, triangle, square, and other geometric shapes always reduces to the number 9.

For Example:

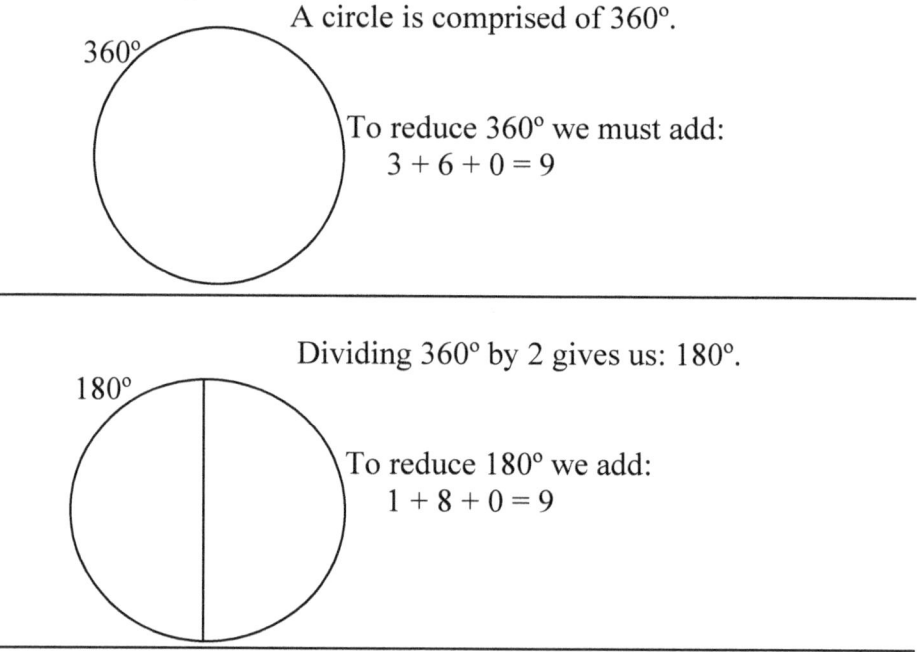

A circle is comprised of 360°.

To reduce 360° we must add:
3 + 6 + 0 = 9

Dividing 360° by 2 gives us: 180°.

To reduce 180° we add:
1 + 8 + 0 = 9

If we were to keep dividing this circle by a factor of 2 this is what we would see as its reduction:

A. 360° = (3+6+0) = 9
B. 180° = (1+8+0) = 9
C. 90° = (9+0) = 9
D. 45° = (4+5) = 9
E. 22.5° = (2+2+5) = 9
F. 11.25° = (1+1+2+5) = 9
G. 5.625° = (5+6+2+5) = 18, (1+8) = 9
H. 2.8125° = (2+8+1+2+5) = 18, (1+8) = 9
I. 1.40625° = (1+4+0+6+2+5) = 18, (1+8) = 9
J. 0.703125° = (0+7+0+3+1+2+5) = 18, (1+8) = 9

We could do this for an infinite amount of time and always get the same digital root of 9.

The inside angles of geometric shapes will also reduce to the number 9.

For example,

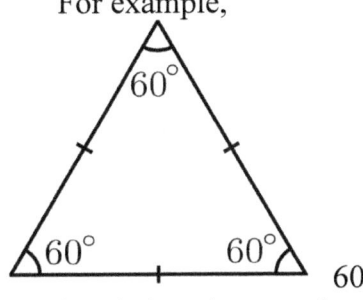

60° times 3 = 180°, 1+8+0 = 9

A triangle has three angles of 60° each.

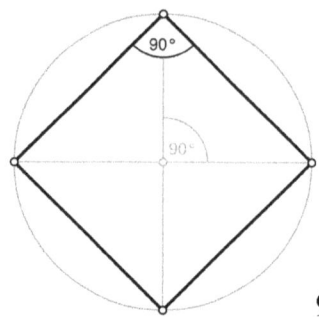

90° times 4 = 360°, 3+6+0 = 9

A square has four corners of 90° each.

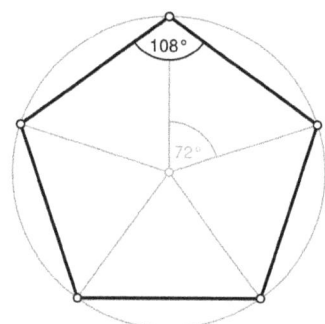

108° times 5 = 540°, 5+4+0 = 9

A pentagon has five angles of 108° each.

A pentagon also has five inner angles of 72° each.

Reducing the inner angles gives us 72° times 5 which = 360° and 3+6+0 = 9

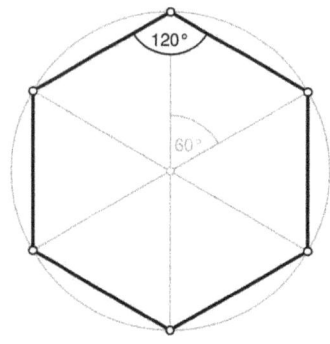

120° times 6 = 720°, 7+2+0 = 9

A hexagon has six angles of 120° each.

A Hexagon also has six inner angles of 60° each.

Reducing the inner angles gives us 60° times 6 which = 360° and 3+6+0 = 9

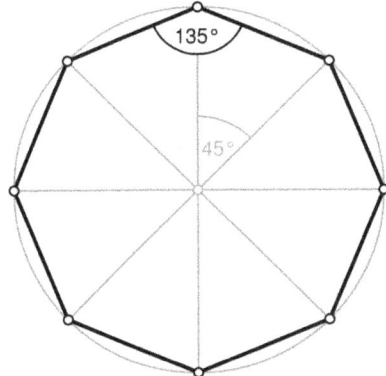

135° times 8 = 1080°, 1+0+8+0 = 9

An octagon has eight angles of 135° each.

An octagon also has eight inner angles of 45° which also reduces to 9.

I could continue this into all of the shapes after this, and still, always the numbers would reduce to the number 9. It makes you wonder if there is something truly special happening around us and within the angles that make all of what we know. And as if knowing about how many times these shapes bring us back to the number 9 let's now explore how sequences of related numbers bring us to that same number 9.

If you were to add all of the Hindu-Arabic numbers (also adding in zero as well) you would get the digital sum of 36, and if you were to reduce 36 you would get the number 9.

$$(0+1+2+3+4+5+6+7+8+9) = 36, (3+6) = 9$$

Paradoxically 9 added to any digit will return the same digit. Watch:
9+5 = 14 (the return) 1+4 = 5
9+7 = 16 (the return) 1+6 = 7
9+4 = 13 (the return) 1+3 = 4

If you were to multiply 9 by any number the resulting sum would always reduce to 9.

Watch:

9 x 5 = 45 (reduce it i.e. 4+5 = 9)
9 x 7 = 63 (reduce it i.e. 6+3 = 9)
9 x 154,795 = 1,393,155 (reduce it)
   1+3+9+3+1+5+5 = 27 (reduce it)
      2+7 = 9

This can be done for infinity. You can literally smash the numeric keypad on your keyboard and multiply that by 9 then reduce it and you will always get 9!

When you look into the realm of time, you realize that there are 1,440 minutes in a day. If you were to reduce this number using the same mathematic process that we have been using, you would get the number 9. There are 86,400 seconds in a day, and this will bring you to a digital sum of 18, which reduces to 9. There are 10,080 minutes in a week which brings us to the reduction of 9. There are 525,600 minutes in a year, and when you reduce this, it becomes the number 9. The amount of minutes or seconds within a day, week, month, or year will always reduce to the number 9.

The 9 code is an inspirational concept in the high science because of how many areas this number is able to hide itself within. It is a number that is located within the reduction of whatever regular geometric shapes angles we look at. It is calculated within the angles of a circle as well as within the realm of time itself. It makes me wonder if there is something more spectacular happening within us and around us that I should be more in tune with. It should do the same for you as well, I don't want you to simply stop looking into this concept after you complete this book but I want you to tell your friends about what I have just imparted on you, tell the young ones so that they may be inspired to want to discover more about these concepts.

Sacred Geometry: The Language of the Creator

Bruce Rawls, the author of the book The Geometry Code, provides a concise definition of what Sacred Geometry is and how sublime of a science it is. He writes, "In nature, we find patterns, designs, and structures from the most minuscule particles, to expressions of life discernible by human eyes, to the greater cosmos. These inevitably follow geometrical archetypes, which reveal to us the nature of each form and its vibrational resonances. They are also symbolic of the underlying metaphysical principle of the inseparable relationship of the part to the whole. It is the principle of oneness underlying all geometry that permeates the architecture of all form in its myriad diversity. This principle of interconnectedness, inseparability and union provides us with a continuous reminder of our relationship to the whole, a blueprint for the mind to the sacred foundation of all things created."[9]

Sacred Geometry is the high science that explores the many patterns utilized in the designing of everything within our reality. It is the realm of high science that looks at geometric shapes and mathematic ratios that can be found within the construction and regulation of the entire universe.

When we humans discovered that on the North Pole of the planet Saturn there was a storm in the shape of a hexagon we were inspired by this proof of geometric archetypes at play to continue investigating everywhere else in the heavens for more examples of this.

---

[9] Rawles, Bruce. "Introduction." The Geometry Code Universal Symbolic Mirrors of Natural Laws Within Us. Accessed January 10, 2015.

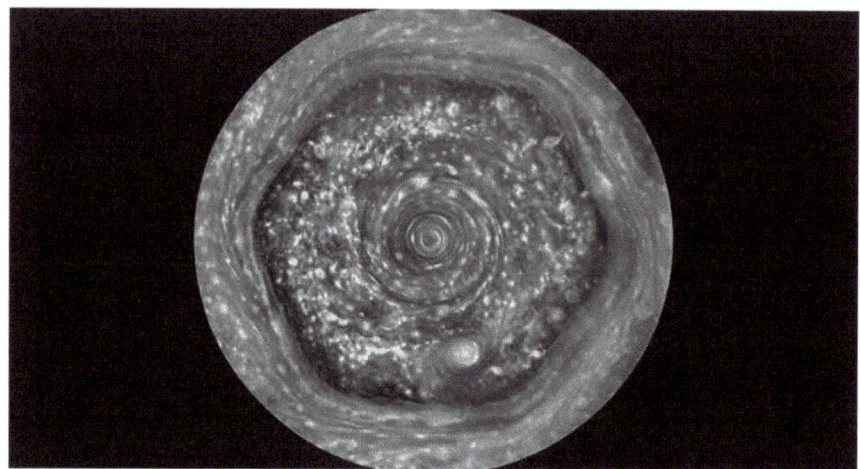

Nasa. "NASA's Cassini Spacecraft Obtains Best Views of Saturn Hexagon." JPL. December 4, 2014. Accessed January 10, 2015.

Public Domain Image by NASA's Hubble Telescope of Saturn in Ultraviolet light.

Byrd, Deborah. "Amazing New Image of Saturn's Puzzling Hexagon | EarthSky.org." Earth-Sky. February 4, 2014. Accessed January 10, 2015

Essence of Astronomy + Astrology:

– Everything is of One; Our Ancestral Roots –

Astronomy, in short, is the study of the cosmos; where planetary objects are, where they are moving, what they are made of, what space is, how space functions, what happens when stars die, and what exactly makes up a black hole or a quasar are just a few concepts that are covered by this field of study. Many ancient societies studied the constellations to predict weather-related events in nature, or for the more esoteric teachings of where life originated, and these are reasons that this field is still so revered to this day. Astronomy helps create a new perspective in all who decide to study it; meaning that it shows us how incredibly small we humans are, how volatile the universe is, how connected we are to each other (by way of how we are made from cosmic star stuff) and it shows us how much we finally know about what we don't know as it relates to that which is above us and within our own planet.

Through the study of astronomy, we have discovered so much about this new world that far out-ages all that we comprehend. On September 5th, 1977 Voyager I was launched into space with the mission of essentially "go play and return whatever you find to us". Its sister space probe Voyager 2 was launched first with a launch date of August 20th, 1977 but Voyager 1's travel speed was so fast that it out traveled Voyager 2 and thus was the first to reach interstellar space; an area of space that we humans have never been to and barely innerstand. The Voyager space probes have recorded (for the first time that we are aware of in this current civilization) the sounds that the planets make! Throughout this Essence I will explore some inspirational concepts uncovered by the study of Astronomy and why it is so important that we never let this knowledge wither away into a forgotten void – we should transcend the human borders that we have drawn on this earth and unify our minds to the greater purpose of cataloging this knowledge and using it for the betterment of a whole life unified with all. With that being said we will now venture into the

Essence of Astronomy starting with the last concept, we visited in the previous Essence; the Hexagonal Storm on Saturn's North Pole.

Astrology on the other hand is involved with finding the potential link between events in the cosmos and how they affect human affairs on Earth. For example, a student of astrology would submit that Comet A has an effect on the spiritual nature of human group A, B, and C because of its connection with zodiac house of Aries. As another example, think of it this way: An student of astrology would believe that the sign Aries is equated with being headstrong, confident, best bonding with the Pisces, and other factors. Aries is a zodiac sign related to astronomical events, societies of antiquity, and creativitiy in explaining the human condition. Whether or not these factors are true is still being resolved, or at least that is my purview on the matter. This essence will focus more so on the Astronomical side but I will not expose which I am referring to as I don't truly see them as being that different because in some areas they seem to have a pretty neat synthesis. For example, the Moon pulls on the tides and this is true, our bodies are also made of over 70% water, so we may be affected by its pull to a certain degree. That deals with both astronomy and astrology to be honest and fair. That being stated, enjoy!

Hexagonal Storm on Saturn

Saturn revealed to us an interesting piece of perpetual geometry that it proudly showcases on its Northern Pole. It is a storm in the shape of a Hexagon that was initially discovered by the Voyager Mission from 1981 to 1982 and later again reconfirmed by the Cassini space mission in 2006. This geometric shape found on Saturn has caused lots of inspiration for all of us who had great reverence for Sacred Geometry, space mechanics, and high science mathematics because we long had adopted the theory that there were some geometric laws prevalent and active in the initial design and management of our cosmos.

NASA/JPL-Caltech/Space Science Institute, Cassini Spacecraft Wide-angle Camera. NASA. July 22, 2013. Accessed January 12, 2015.

Shown in the photograph above is the picture from the Cassini Spacecraft as it observed the planet Saturn. This picture is in true color [*true color essentially means the natural color of a defined object*] and shows us the hexagonal storm (or vortex) on the North Pole of Saturn. Here you have a geometric shape, which we can make using mathematics (utilizing the current Hindu-Arabic numeral scale) shown in design and management of the universe.

Dunbar, Brian. NASA. April 29, 2013. Accessed January 12, 2015 http://www.nasa.gov/mission_pages/cassini/whycassini/cassini20130429.html.

    In the picture above provided by NASA, we can see the false color image of that same hexagonal storm (hurricane) on Saturn's North Pole. I have included this picture in this book because I want to take a second and reflect on how hurricanes follow the Fibonacci sequence and how in this picture we see the Fibonacci sequence plus the hexagram on a massive planet in the shape of a sphere. These are all some high science concepts happening in the cosmos that we should be inspired by to look further into their origins and meanings. It is because of these interconnected concepts that I consider the Hexagonal Storm on Saturn to be an inspirational concept in the high sciences. It is because of how sublime of a revelation it was to find out that Saturn was holding such a precious gem within its contents, that geometry is at work in the design and management of the cosmic bodies, and that there can be such beauty in the innerstanding of how majestic our universe is.

Observational Universe

Open your minds. Open them in preparation for this grand journey you are about to take into the cosmos through thought. Open your minds and allow the sacred energy of your imagination in pensive thought to bring you to the awesome realization that is the scale of the heavens around us.

{"We have developed from the geocentric cosmologies of Ptolemy and his forebears, through the heliocentric cosmology of Copernicus and Galileo, to the modern picture in which the earth is a medium-sized planet orbiting around an average star in the outer suburbs of an ordinary spiral galaxy, which is itself only one of about a million million galaxies in the observable universe."

– Stephen Hawking, *A Brief History of Time*}

There is an experiment that I call "The How Far Thought Projection Experience" that will take you on a journey throughout space in using your imagination in pensive thought to discover the scale of it all. I ask that in this point of the book you seriously perform this experiment and take part in what purpose for it I have created.

Proceed to the next page when you ready.

Experiment: The How Far Thought Projection

<u>Step 1</u>: On a clear sky and starry night preferably with a bright whole-moon I want you sit in the open lotus meditation pose and gaze upon the twinkling stars and the moon; wholly ignore all else moving around you on the earthly plane. I want you to realize how far away the actual stars are, that their light takes light years just to reach your eyes and that the light you're bearing witness to from the stars may be the ghostly remanence of a star now gone.

Night Sky with Moon and Stars by George Hodan

Step 2: I want you to astral project yourself into space as if you were looking at the earth from a spaceship. Realize at this moment where on this planet you reside and how small you are from this scale. Innerstand that from this distance your problems, joys, and all else no longer matters.

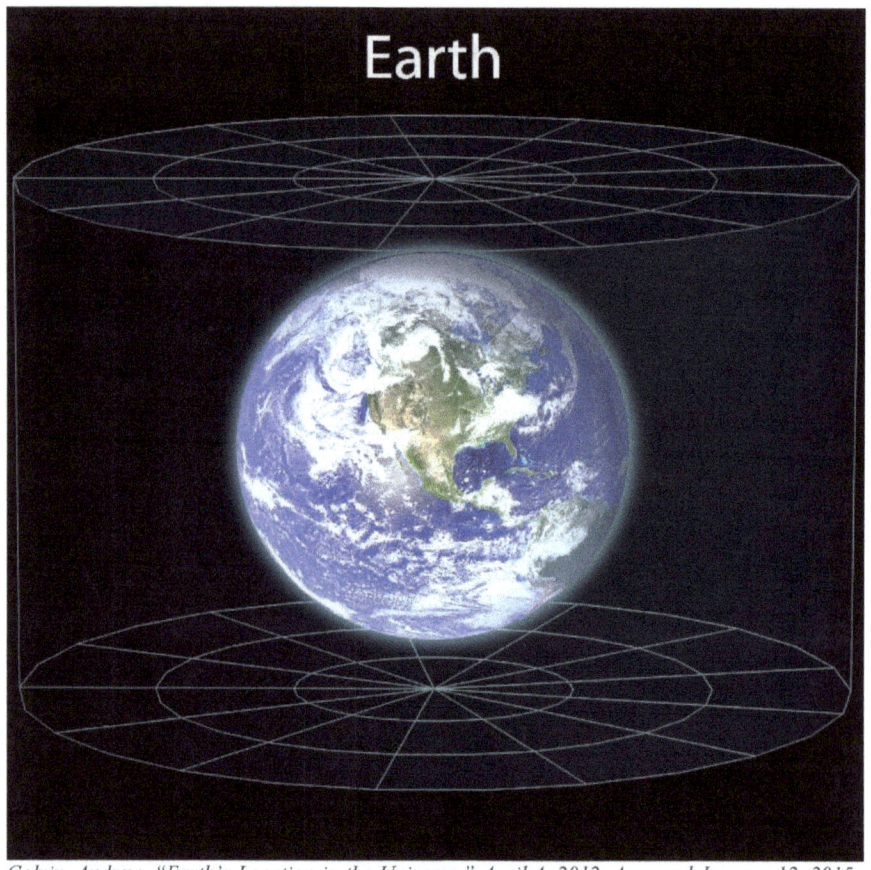

Colvin, Andrew. "Earth's Location in the Universe." April 4, 2012. Accessed January 12, 2015. https://commons.wikimedia.org/wiki/File:Earth's_Location_in_the_Universe_(JPEG).jpg.

Step 3: Realize that Earth is but a small planet within our immediate solar system and that we revolve around our Sun in a cadence with many other planets, comets, and the Kuiper Belt. Take a second to see where exactly our planet Earth falls within this cadence and remember that.

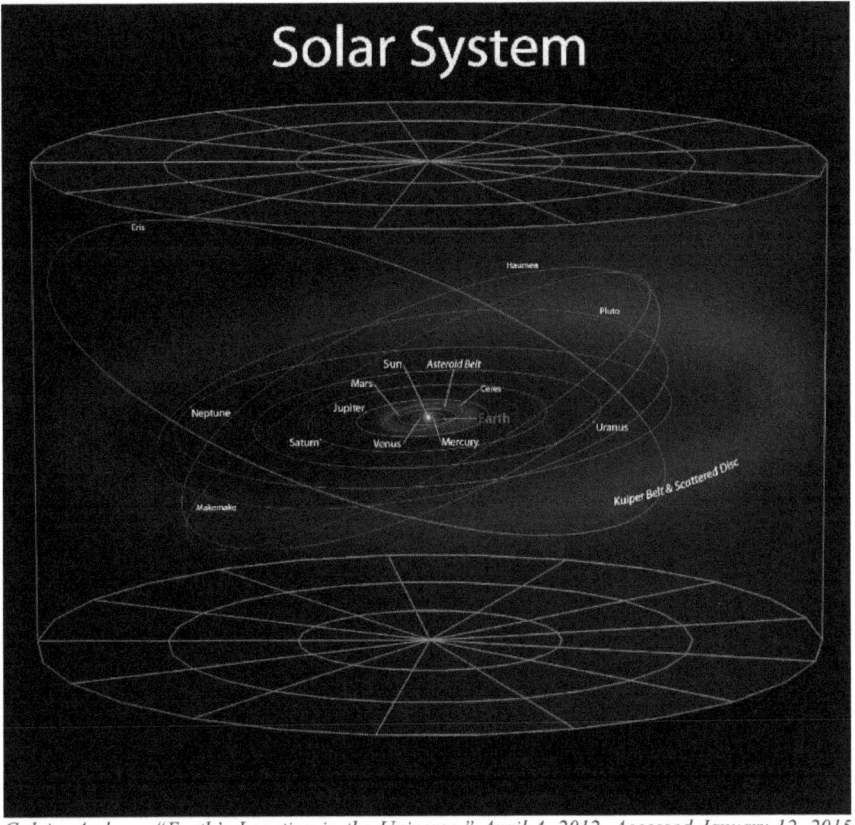

Colvin, Andrew. "Earth's Location in the Universe." April 4, 2012. Accessed January 12, 2015. https://commons.wikimedia.org/wiki/File:Earth's_Location_in_the_Universe_(JPEG).jpg.

Step 4: I want you to realize what outside of our solar system looks like and how many other solar systems are now in our current view from just outside of our current solar system. I want you to take a minute and look around at the other solar systems around us, then realize where we fall within this neighborhood.

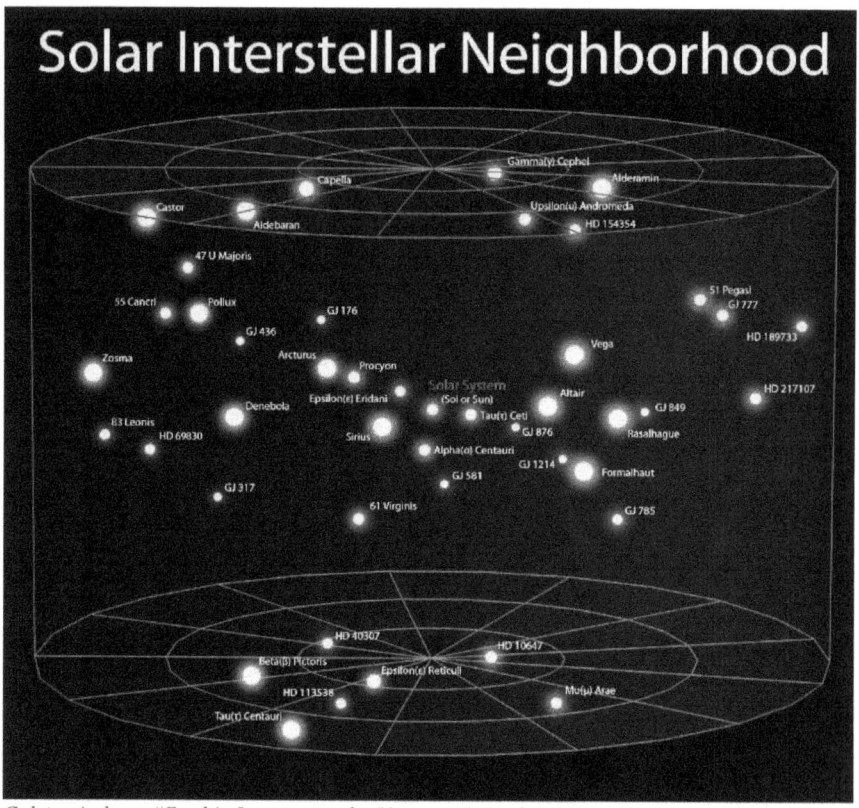

Colvin, Andrew. "Earth's Location in the Universe." April 4, 2012. Accessed January 12, 2015. https://commons.wikimedia.org/wiki/File:Earth's_Location_in_the_Universe_(JPEG).jpg.

Step 5: Astral project yourself outside of this neighborhood and you'll find yourself now outside of our galaxy. You are currently looking down upon our own Milky Way Galaxy and just look at how small earth falls within this system. Take it all in and be ready because we are about to go much further out into the cosmos.

Colvin, Andrew. "Earth's Location in the Universe." April 4, 2012. Accessed January 12, 2015. https://commons.wikimedia.org/wiki/File:Earth's_Location_in_the_Universe_(JPEG).jpg.

Step 6: Astral project yourself far from our own Galaxy and gaze upon all of the other aspects of what is around us that we are aware of. Take it all in always keeping in the back of your mind how far from earth we have now traveled outward into the cosmos.

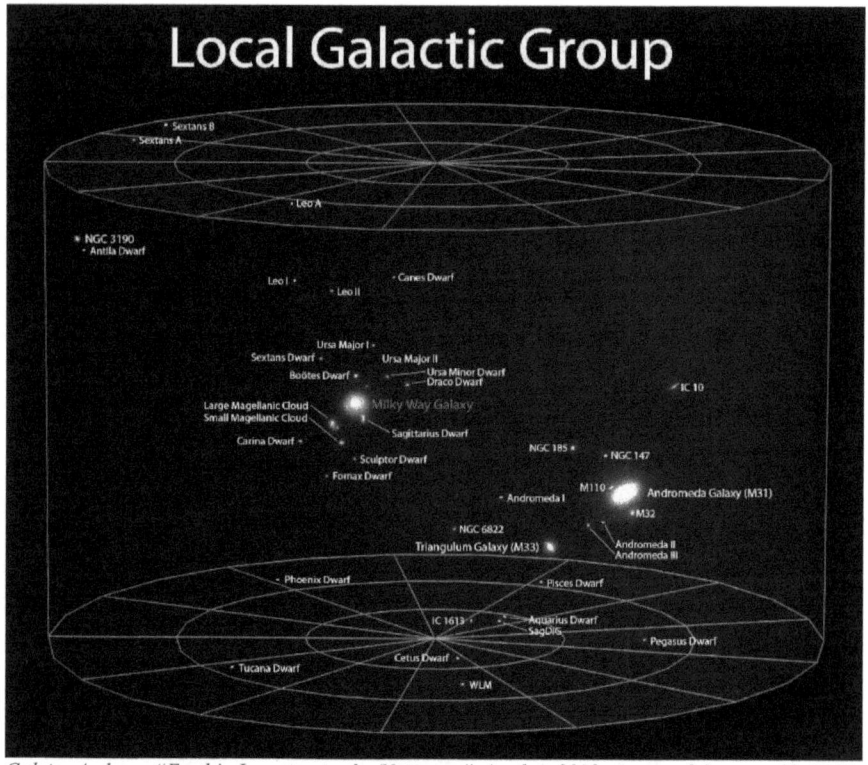

Colvin, Andrew. "Earth's Location in the Universe." April 4, 2012. Accessed January 12, 2015. https://commons.wikimedia.org/wiki/File:Earth's_Location_in_the_Universe_(JPEG).jpg.

Step 7: Astral project yourself yet again outward from our Local Galactic Group and you'll now find yourself within what scientists call the "Virgo Supercluster" which is essentially a grouping of other Local Galactic Groups that all hold galaxies and solar systems, and planets within their respective systems. Take it all in and realize from this scale how far we have traveled from our home planet of Earth.

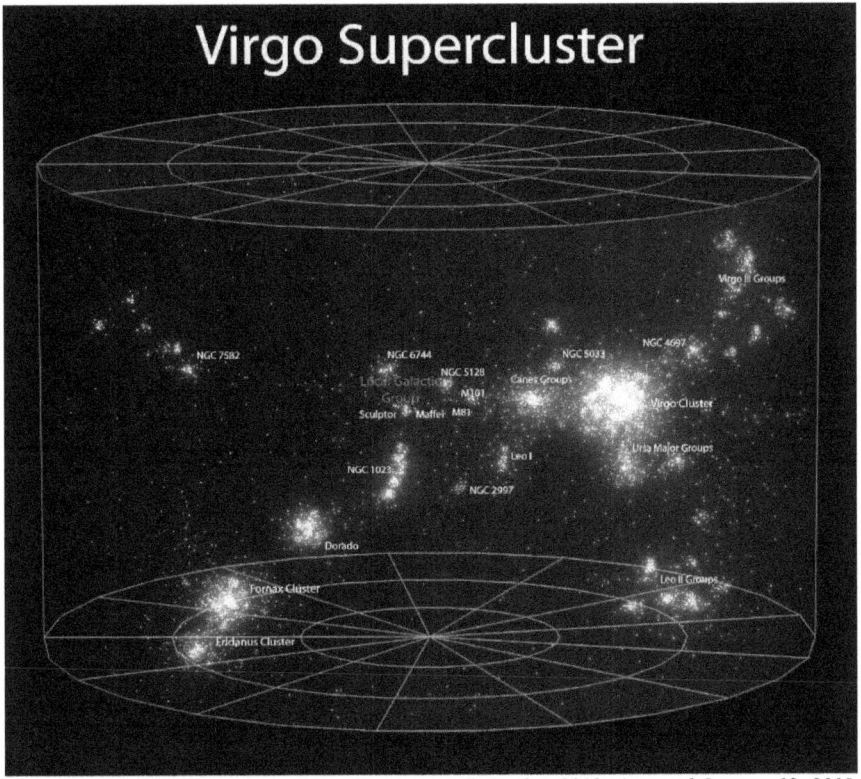

Colvin, Andrew. "Earth's Location in the Universe." April 4, 2012. Accessed January 12, 2015. https://commons.wikimedia.org/wiki/File:Earth's_Location_in_the_Universe_(JPEG).jpg.

Step 8: Astral Project yourself once more and you'll find yourself within what scientists call the "Local Superclusters" which contain more superclusters, galaxies, planets, suns, solar systems, and moons. Realize from this massive scale how small we are and realize that we still are not completed with our journey into the cosmos.

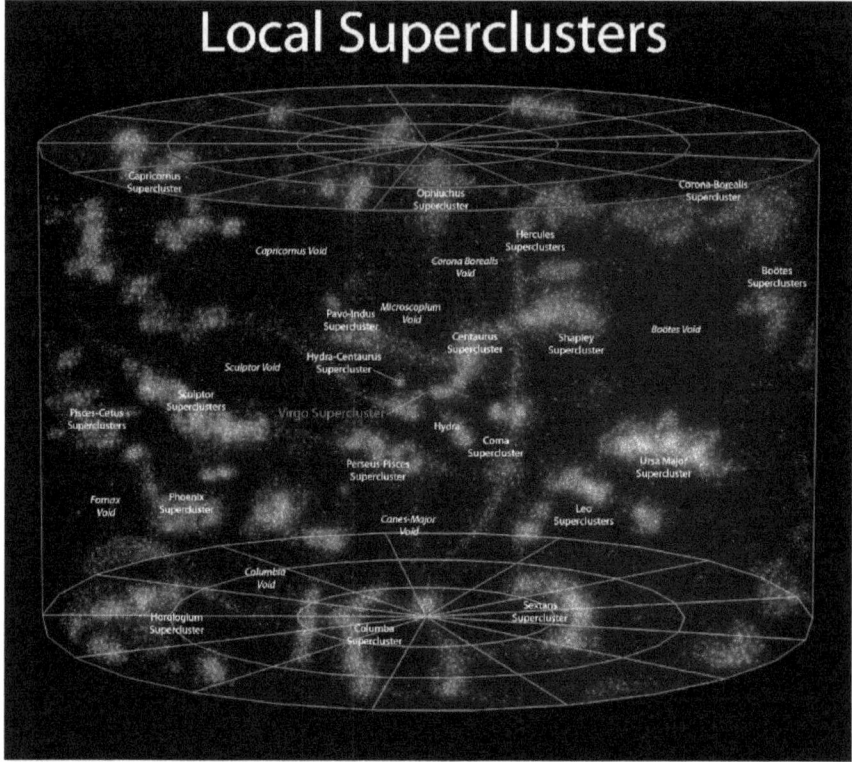

Colvin, Andrew. "Earth's Location in the Universe." April 4, 2012. Accessed January 12, 2015. https://commons.wikimedia.org/wiki/File:Earth's_Location_in_the_Universe_(JPEG).jpg.

Step 9: Astral Project yourself one final time and you'll find yourself in what we call the "Observable Universe." This is all that we know of currently. We have no idea what else is out there and yet when we look at how far we've traveled we have come so far, and we can now see how small we are.

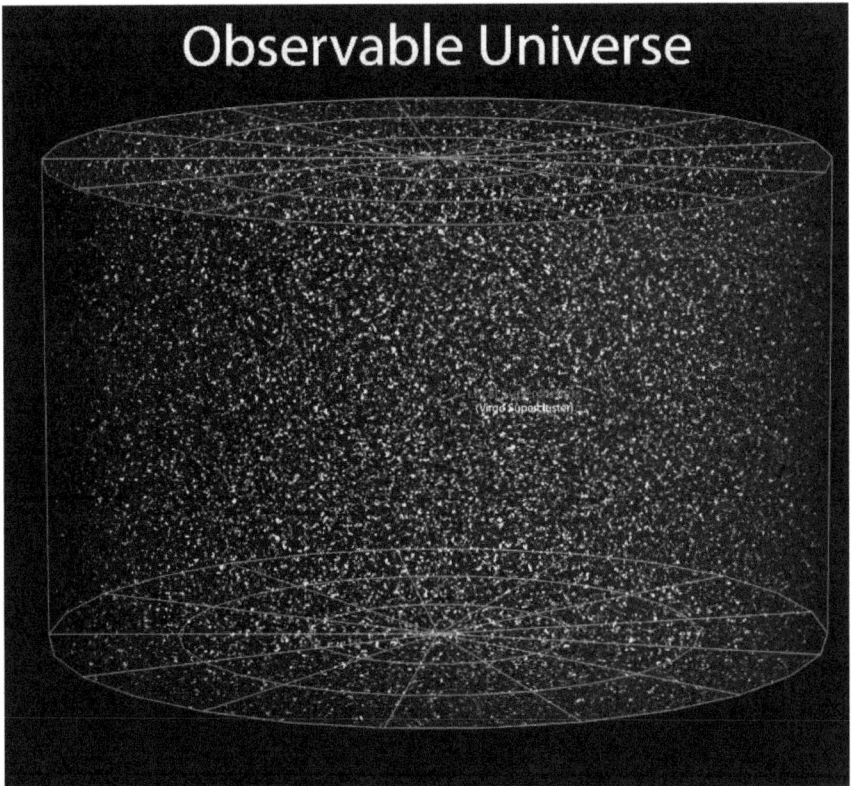

Colvin, Andrew. "Earth's Location in the Universe." April 4, 2012. Accessed January 12, 2015. https://commons.wikimedia.org/wiki/File:Earth's_Location_in_the_Universe_(JPEG).jpg.

Planetary Sounds

That's right the planets are alive and making all sorts of strange noises. Voyager 1 and Voyager 2 recorded the electronic vibrations being emitted from the various planets, rings, moons, and the electromagnetic fields of planetary objects. They also recorded the planetary magnetosphere, the solar winds, charged atomic particle interactions of the planets, and trapped radio waves being emitted between the planets and the inner surfaces of their respective atmospheres. Their current website archive of these encounters can be found at www-pw.physics.uiowa.edu/voyager-audio.html, and it is quite the audio adventure.

In addition to that unaltered raw data NASA has also presented a CD called "Symphonies of the Planets."

They are worth listening to, if not for the sake of curiosity alone then for the sublime fact that you're bearing witness to how the planets sound!

Zodiacs, Constellations, Ages & Cycles in the Heavens

The ancient Africans were responsible for mapping out the heavens and their movements across the universe, which is why they are credited with calculating and showing in their monuments, arts, language, and life what modern Western scientists refer to as "the Great Year." Such proof is found in artifacts left behind by the Kemites is in the form of what is called the "Dendera Zodiac" which was located originally (before it was cut, sawed, and blown out of its original home to the Musee du Louvre Museum in Paris, France) in the Hathor Temple at Dendera. Many scholars and so-called scholars debate the age of the artifact where one side argues that it is tens of thousands of years old and some arguing that was established after Kemet fell to other societies. Given the research and evidence from Kemet, contributions by Master Teachers such as Dr. John Henrik Clarke, Dr. A.A. Yosef ben-Jochannan, Anthony Browder, Ashra Kwesi, Dr. Cheikh Anta Diop, and many others I am more inclined to submit that this knowledge was around long before Kemet fell to invading forces. It is highlighted in the many other aspects of their monuments that date all the way back to Kemet's 1st Dynasty, and even including knowledge established in Nubia long before Kemet was a civilization. Common sense will even show you that in order for the Kemites to have known when the planting season is, when the harvesting season is, when the flood of the Nile River would be, and the affairs of the general weather patterns they would have had to known how to read the heavens and track the patterns which takes enormous time, patience, and intelligence.

In the book The Meaning of Hotep: A Nubian Study Guide, written by Anpu Unnefer Amen he states that "Just as there are twelve zodiac signs that complete their cycle in harmonious balance with our earthly year, there are also twelve zodiac signs that complete their cycle in harmonious balance with the Great Year. Each of these signs represents an "Age," and each sign stays in its "House" where it can be observed telescopically for a period of 2,160 years. As one Age transitions to the next, it does so "backward" in relation to our yearly zodiac. So instead of shifting from Cancer to Leo to Virgo, it would move from Virgo to Leo to Cancer, etc. The Completion of the procession of the Twelve Ages marks a Great Year (25,920 years). Our planet (Earth) is also in harmony with the Great Year. The earth is tilted on a 23 ½ degree angle due north and does not spin in a perfect circle. It can be said to "wobble" sort of like a spinning top. As a direct result of this "wobble effect," the planet takes 25,921 years to make a complete gyration on its axis. The cycle completes itself in conjunction with that of the Great year."

---

[10] The Dendera Zodiac, photo taken by: Chatsam on November 3, 2013. Found in Wiki Commons

The ancient Africans are the reason that we in this modern era are able to know what we know about the heavens. In fact, there was another monument built by the ancient Africans of Kemet at Abu Simbel. It is such a profound place that after the music group Earth, Wind, and Fire visited it they pretty much dedicated the entire album "All in All" to it, placed Abu Simbel on their album cover, and I even noticed that their music changed to contain more references to light, positivity, joy, history, and personal development (songs like Pure Gold, Devotion, etc.)

The Abu Simbel Temples are two gargantuan stone temples located on the west bank of the Nile River in a place we now call Abu Simbel, in southern Kemet. They were built by (and I will conversationally use this Greek term but as a disclosure, the Kemites did not refer to their leaders as "Pharaohs") Pharaoh Ramesses the 2nd (1303-1213 B.C.E.) It was designed, constructed, and aligned in such a way that on two days of the year (October 22, and February 22) the sunlight of the day would shine brightly into the "inner sanctuary and light up threes statues seated on a bench, including one of the pharaoh."[11]

---

[11] Jarus, Owen. "Abu Simbel: Temples of Ramesses II." Live Science. June 11, 2013. Accessed January 16, 2015.

12

In order for the ancients to showcase the sun hitting that innermost room twice a year they would have had to have known that the earth rotates on its axis by 23 ½ degrees and orbits the sun instead of the sun orbiting the earth and the earth being flat as was taught by Catholic churches from their early, and ignorant of astrology days.

It is also worth noting how some cultures disrespect even our most prized accomplishments. In the 1960's the current inhabitants of the land of Kemet decided to move the temple because they wanted to make a High Dam. Think about this for a second, the original builders and astrologers that decided on where to place this monument picked out this specific area for specific reasons which have now been violated like building a theme park over an ancient Native American burial ground. The place they moved it is a desert plateau and the original location of where the temples at Abu Simbel were is now flooded due to the dam. It has been stated that "everything looks just as it did before; it is enough to make one doubt that the temples were

---

[12] Credit: WitR | Shutterstock – Jarus, Owen. "Abu Simbel: Temples of Ramesses II." Live Science. June 11, 2013. Accessed January 16, 2015

moved at all."[13] So the original builders' intent, plan, and design have been modified and somehow, to many non-natives of the land of Kemet, it is as if they weren't moved at all. This mindset completely ignores the astrological science that went into its original placement and violated the sanctity of respect that should be in place for the people that gave the world much of its knowledge.

Putting this sad story into the memory banks of history, the Zodiac, Constellations, Ages, and Cycles in the heavens are inspirational concepts in the high sciences due to the enormous amount of knowledge they withhold; much of which is still prevalent and accurate to this day. For it was not only the ancient Kemites, but also the Maya, and many other ancient cultures that great reverence to the inner workings of the universe that we all reside in and are so governed by. This knowledge is what guided Benjamin Banneker [*a relative of the Dogon Tribe of West Mali*] in creating the farmer's Almanac in America that taught people about the planting and harvesting seasons. [*Given this man's many significant accomplishments for the United States of America it is quite horrible that all he received post mortem is Banneker Circle in Washington, D.C.; a small and un-lauded place of nothingness.*] Their studies of these heavenly phenomena completely and wholly affected how they lived much of their lives, and how they found balance. Many of the true and sincere scholars of the past days (and modern days) show the same level of respect and adoration amongst each other. Meaning, that in the heavens they see themselves, in each other they see themselves, in the grass they see themselves, in the rocks they see themselves, and it is through that innernstanding of ancient astrological sciences (and modern astrological studies) that all of our lives are positively affected. Because of Neil DeGrasse Tyson, Bill Nye (The Science Guy) and many other unnamed scientists I have been positively impacted in a way that completely changed how I function on this planet. I see myself in every aspect of all and in that I realize that no matter what political or cultural differences that at the core we are all cosmic star stuff from origins unknown and that we should be

---

[13] Robert Morkot in an article in the "Oxford Encyclopedia of Ancient Egypt" (2001, Oxford University Press).

on the sublime and everlasting journey of cataloging true knowledge uncorrupted by racism, bias, favoritism, envy, disrespect, or other negative factors.

The concepts of the Zodiac signs, the Great Year calculations, the Cycles of the Heavens, and the Ages are critical to understanding where religious stories come from, what they mean beyond the literalistic readings of them, how this knowledge helps us better live our lives, and how to better work with one another to find our true roots.

The more you allow your mind to wander as it relates to these this subject matter the more information you'll find, the more information you'll share, the more conversations (and even arguments) you'll have, the more we will all grow, and the more able we (as a globally integrated) peoples will be able to "utilize our ears function as a garbage filter"[14]

We as "the one cosmologically united in molecular origins family" [yes, feel free to reuse any of my sayings in your daily life] must be able to accept what has happened throughout history. The first step to resolving a problem is admitting that you have one and whether by unknown ignorance of fact or willful ignorance of fact many people aren't able to see that we have a problem. Remember what I told you about what Count Volney said as he overlooked the ruins of Kemet? In the late 1700's the Atlantic Slave Trade was in full force, and it was seen as a good institution by the European powers of the day. In recap, Volney essentially stated (as he overlooked the fallen and ruined lands of Kemet) that how is it a reality that the very people that gave the world literacy, the sciences, astrology, mathematics, law, civilization, morality and so much more have been reduced to subhuman servitude simply because of their black skin and curly hair? He stated this with such disgust because he was able to see past the ignorance of fact that afflicted his European brethren. Volney was able to innerstand that he acquired his intelligence from ancient societies such as the Africans, and he was at peace with that. The problem in the modern era is that many people don't want to accept

---

[14] J-Live's song called "School's In"

this for whatever illogical reason plagues their mind. Once humanity accepts that the early Europeans messed up and were wrong in what they did to the Africans – Once humanity acknowledges that Africans contributed heavily into the world as we currently enjoy – and they can do this without malice or a feeling of superiority (or inferiority) then humanity can begin to realize that we are one. Astrology alone can help this process, and that is in large part what these inspirational concepts are meant to promote. Be at peace with where we really come from and who has really contributed what and when. We MUST learn to put aside illogical political motives, extreme nationalism, disrespect towards cultures (ancient and modern), and the façade of educational (institutional) lies being taught as reality.

All of the great minds that we revere were able to innerstand this. Nikola Tesla, Albert Einstein, and so many others; internalize these inspirational concepts and truly allow them to modify your perspectives. Realize that at our molecular cores we, whether black, white, yellow, fat, skinny, old, or young are all made from cosmic star stuff and that in a true sense we don't know much else about where we come from; most likely because of illogical divisions of land in a world that no person or nation owns – We are all one, and our minds should be in true respect of that; after the ills of the past have been addressed and properly resolved.

In the spirit of this new perspective that I know you are seeing I implore you to research more about the origins of the mechanics that so govern our time today. As an example if you were to investigate why the twelve months have a mixture of 28, 29, 30, and 31 days you'll wind up finding that they should all have 30 day cycles but that certain rulers of early nations used to borrow days from the other months to extend reverence for themselves – here we are all these years later, and we still use that strange method of calculations out of habit and blind tradition. "Perspectives change when knowledge is obtained". Remember when I stated that? How true does it stand in this instance right here? I know your perspectives just changed from that small nugget of little-known information about the systems in place that regulate our concept of time and order, and hopefully you ask questions on everything. Why celebrate this? Why do we do that? What does this mean? Etc. These are the questions.

{"In the end, there is only race: the human"
– George Edward Moore}

Great Red Spot; Jupiter

When you think of the planet Jupiter what usually comes to mind is one of its most signature planetary features; its Great Red Spot. Its most iconic feature is that perpetual storm (Great Red Spot) that has baffled planetary explorers for a mighty long time. The questions of what is it, where did it start, what started it, why is it the color it is, and what does it mean have been asked, explored, asked again, and researched all in hopes of finding some solid answers that allude to the inner workings and origins of the storm. What we learned out is that the great red spot is rather enormous in scale, like so enormous that it is bigger than the entire Earth plus some. We have also found out that it is shrinking slowly, or that at the very least is in a stage of regression – possibly even a cycle where it grows and shrinks.

15

Through human curiosity, the spacecraft called Cassini launched, in conjuncture with NASA has captured some incredibly vivid and high-resolution (close-up) photos of this storm. It is because

---

[15] NASA's Cassini spacecraft caught this image while it was on its way to Saturn on December 29, 2000. NASA/JPL/Space Science Institute.

of this curiosity and exploration of the heavenly bodies that I refer to the Great Red Spot on Jupiter as an inspirational concept in the high sciences because it showcases (yet again) how much we know about how much we don't know. It also solidifies the sheer size of the heavenly bodies that orbit our average sized sun, in a solar system that orbits (along with billions of other solar systems) a supermassive black hole (the Milky Way Galaxy) which is not even the end of space but only what we are able to immediately (through spacecraft) explore.

Jupiter also has an interesting feature where it has bands of its atmosphere that spin in opposite directions. Some going leftward in motion and some going rightward in motion; some also rotate faster and some rotate slower – with that Great Red Spot in the middle of it all.

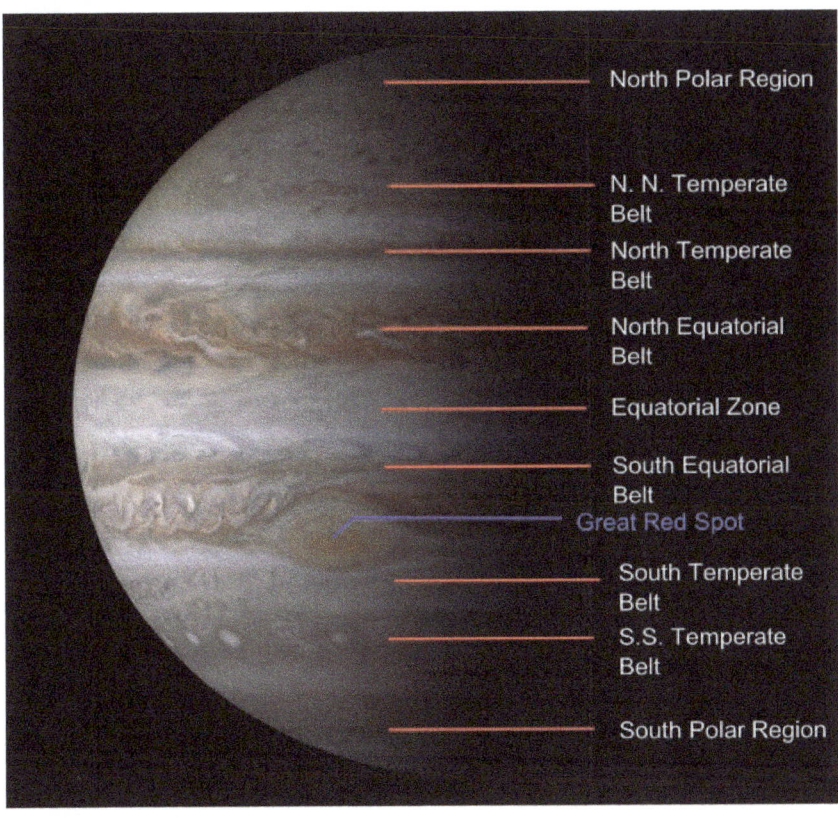

[16] Jet Propulsion Laboratory / NASA. Cassini Spacecraft. PIA04866

This is an inspirational concept in the high sciences because firstly, had it not been for Jupiter many of the asteroids that could have obliterated Earth were absorbed by Jupiter and instead collided with that planet – so, in essence, we owe Jupiter a great deal of thanks. Secondly, here we are in the year 2015, and we haven't cracked the inner mechanics of that planet, nor have we ever been able to collect a sample of what it is made out of, nor have we (for that matter) ever been to interstellar space (until now thanks to Voyager 1 from 1977). We really don't even innerstand why it spins with different directions, and what that red spots story is. We should be inspired to investigate with, true sincerity the inner mechanics of the universe.

Black Holes

A black hole is an area in space where gravity's pull is so intense that matter, including light, cannot escape. They are usually formed when a star (like a sun) collapses in on itself and its matter thence becomes super condensed to an intensely small area of space and from that collapse the gravitational pull of what was once a planetary body begins to absorb everything around it as it gathers more and more matter within its center.

[17] Alain R. – Simulated Black Hole – Wikimedia Commons. September 8, 2006

The question of what exactly is in a black hole is just as much a mystery now as it likely was in ancient times. Black holes have something that scientist refer to as an "Event Horizon" which is the point before the event that we are unable to innerstand. Meaning, we literally have nothing but theories as to what is happening after you fall into a black hole. Why is this? Because even if we were to send a space probe into the center of the Milky Way's Supermassive Black Hole the space craft would just vanish and it could only communicate to us everything up unto the Event Horizon.

We have theories for days on what many people believe is within the grasps of a black hole and scientist are now beginning to start creating super mini ones in labs but as of this current juncture in 21st-century human history, we haven't the faintest clue what lies beyond the Event Horizon. You may ask yourself this is interesting but why does it even matter and why is it an inspirational concept in the high sciences? Well, that's an excellent question to which I answer what is in the center of our galaxy? A supermassive black hole. What do black holes usually do? Absorb matter indiscriminately, builds universes, and tear old ones apart. For all we know a black hole could be a wormhole [*a space-time anomaly where time and space are distorted. Think hyper speed travel from those sci-fi shows that always show a spaceship entering into some space hole that spits them out somewhere else in space; that's essentially a wormhole*], inside of a black hole time might not even exist as black holes violates almost every law of physics that we are able to innerstand. You see, this matters because by asking these questions we explore, and in that exploration, we are able to know more about the happenings around us. We cannot just go our whole lives ignorant of the world around us; well you could, but you'd be a knowledge void hermit, and that's no fun.

By innerstanding what black holes are and how they work we increase our sublime consciousness; our power. Through researching concepts such as this we have re-discovered ancient knowledge – the fact that we are as large to the universe as ants are to earth – and this has completely changed the perspective of all who venture into this type of knowledge. Imagine if we were to finally discover what is after the Event Horizon [*now that you have a general idea of what it is*] – imagine how much more our perspectives will change. Not only that but the more we study ancient societies of times past we learn how in tune they were with this knowledge and we are dumbfounded at how brilliant and meticulous they were with cataloging it all. While we can't go back in time to speak to the ancient Kemites of North Africa, or the ancient Maya, or the ancient Sumerians and such, there is one tribe in particular that was in tune with this knowledge. That tribe is known as the Dogon Tribe of West Mali, and they are believed to be the descendants of the Kemetic civilization that migrated away from the invaders and turmoil that obliterated their homeland. This tribe (before the invention of telescopes) knew about the rings of Saturn, moons of Jupiter, Sirius A, and Sirius B. It was in 1946 that a French anthropologist by the name of Marcel Griaule studied the Dogon for thirty-three days and was astonished at this Sudanese astrological systems knowledge. They were confused as to how the Dogon could have possessed all of this knowledge that modern science was just starting to discover – when I first began researching the Dogon Tribe of West Mali I wasn't at all confused as to how they knew all of this because while I don't completely innerstand how they acquired this knowledge I know that ancient societies in Africa were far beyond where we are in so many realms of study and because of invading forces in Kemet, Mali, the Congo, Somalia, Ethiopia, and the Atlantic Slave Trade much of that knowledge was destroyed, corrupted, stolen, or misattributed by racist men of the times. Just because a people are seemingly basic (natural) doesn't mean that they aren't highly sophisticated, and this is something that people of this modern 21st century need to innerstand as well, but I digress.

"The problem of knowing how, with no instruments at their disposal, men could know the movements and certain characteristics of virtually invisible stars has not been settled, nor even posed."
– M. Griaule, G. Dieterlen, 'A Sudanese Sirius System', ibid,p.59

I will revisit the Dogon Tribe after we explore Black Holes and Quasars a little deeper – I feel that the Dogon Tribe deserve much more reverence than can be contained in the sub-chapter of a chapter.

Black holes are an inspirational concept in the high sciences because of how critical they are to holding together and creating the heavenly bodies. Black holes are creators, black holes are destroyers, black holes are continually transferring objects energy and matter, and yet we know so little about them. We barely catch them on camera, and we are not making enough of a unified effort to better catalog knowledge about this object in the heavenly cosmos. I challenge you, my reader, to ask and answer these questions throughout your days on this earthly plane:

- What is in a black hole?
- What happens if I were to travel into one?
- Do black holes really distort time?
- What exactly is time?
- What (if anything) can escape the grasp of a black hole?
- What happens if our Galaxy's black hole spits out all of what is has absorbed over the years?

Take humanity further into the light of knowledge and the warmth of her embrace. What we don't know we should know or be working to know. We should embrace our unified minds behind the cataloging of knowledge of the universe with proper and unbiased attribution to all parties involved in the discovery of that knowledge.

{"Seek the answers and enjoy the journey."

– Heh Heru}

Quasars

Quasars are... well... no one really knows for certain what they actually are [*just as strange and hard to decode as black holes*] but there is a pretty strong theory that describes their formation and mechanics. That theory points to what happens when supermassive black holes are gorging themselves in the matter of the former planets, stars, and other universe matter. As supermassive black holes absorb that matter, they absorb it in a spinning disk or circular format. This absorption process speeds up the matter caught within the black hole at a rate that constantly increases. This makes the matter rub up against other particles of matter which in turn causes friction – friction leads to heat – and that heat leads to light being a byproduct which we are able to see. That output of intense light is what makes these quasars so ultra-bright within the universe and usually able to outshine entire host galaxies.

The first quasar identified in the (modern era) universe is called 3C 273 and is located within the constellation Virgo (the Virgin) and was identified by T. Matthews and A. Sandage in the 1960's. This was the point in our relatively young exploration of the universes mechanics and it was yet another seed in the ground that scientists began to grow in the vast garden that grows ever more with each discovery.

In essence, we simply found a seed, planted it, watered it, and gave it sunlight but had (and still have) no idea about what exactly we were growing. It is because of curiosity like this that we are able to grow in our collective mental capacities as it relates to the many layers of the universe – the 1960's gave us the first known image of a Quasar and more recently the Hubble telescope captured the best known image of this magnificent light – who knows what the future holds. We build upon the layers of knowledge from the past and through respect and reverence for those who come before us in the true spirit of discovery for humanity, and we grow as a human race because of that.

The discovery of quasars was another instance where we [humankind] realized just how small we are in the ever growing scale of the universe. We make the common perspectives failure in thinking that we are so massive, the world revolves around us, and that we can own things in this universe when in all actuality the reality is that it is the nature of the universe to reject ownership labels and illusions of power and dominion – we are simply at the whim of the machine that is the universe. We are infants in our knowledge and should be inspired to learn more.

[18] NASA. "NASA's Hubble Gets the Best Image of Bright Quasar 3C 273." NASA. Accessed January 19, 205.

{"Twinkle, twinkle, quasi-star
Biggest puzzle from afar
How unlike the other ones
Brighter than a billion suns
Twinkle, twinkle, quasi-star
How I wonder what you are.

– George Gamow}[19]

---

[19] Originally published in *Newsweek*, 25 May 1964. In Ivor Robinson, Alfred Schild, and E. L. Schuckling (eds.), *Quasi-Stellar Sources and Gravitational Collapse* (1965), 472.

Speaking esoterically here, no one really knows if the ancient societies, that were so great at perfectly predicting planetary events were aware of these giant flashlights in the universe. One would have to assume that they were most likely in tune with such events given the technical accomplishments that we can't reproduce today. An example of this is with the Great Temple at Giza, here we have a structure built around 4,600 years ago, that is supermassive in scale, an eight sided figure, pointed to magnetic north (within $3/60^{th}$ of a degree error, making it the most accurately aligned structure on Earth to date), constructed of multiple types of stones, aligned with planetary objects, and encoded with such equations as the mathematical speed of light, and the golden ratio – this is just me speaking here but I know that the ancients must've known about these and I only wish we were able to reach back into history and ask them what they knew.

And it wasn't only the Kemites of North Africa that I believe had access to this knowledge but also the ancient Incans that built Machu Picchu, the builders of Stonehenge (a prehistoric monument structure in Wiltshire, England), the Dogon People of West Mali, and the ancient Nubian/Kemetic Natives that built the astrological site called Napta Playa. There is so much evidence that we are uncovering that showcases just how highly advanced these ancient builders were, and perhaps it was because of the types of alternate lifestyles they lived. Alternate being that they didn't have to worry about bills, debts, or other modern stresses that societies within the social contract have to deal with on the daily basis – almost a monk-like existence is what many of these cultures had and so much of their time was likely devoted to learning more about the heavens – they didn't have video games, or movie theaters, or other distractions or pleasures – call them what you will as they are perspectives based but I digress.

{"Throw your dreams into space like a kite, and you do not know what it will bring back, a new life, a new friend, a new love, a new country."
– Anais Nin}

Quasars are inspirational concepts in the high sciences because of their mysterious nature, our inability to visit one and learn about its mechanics, our lack of innerstanding as to how they act as Galaxy creators or eaters, and how much power they hold within them. We should be inspired to learn more because of how small and powerless we are to them. Think of it this way if I am ever put in a position where I must play and win a chess game for the survival of humanity then it would help if I've studied the game of chess prior to me needing the knowledge of how to play it properly correct? Remember the movie Armageddon with Bruce Willis? Well if you do then you'll remember how much science and planning went into their mission that was critical to the survival of the entire human race. It was incredibly beneficial to humanity that they had subject matter experts on call to help address the looming destruction, wasn't it? That example may be extreme and unlikely to a person without the innerstanding of just how active our universe is and how many times we have come close to utter annihilation, but it is in our best interest to study these concepts and to inspire others on why these are so important and cool for that matter. I mean how cool is it that a Quasar light years away [a light year is almost six trillion miles – yes, that's 6,000,000,000,000 miles away.] is able to be seen by our telescopes. How cool is it that this distance is so minute and but a mere half of a half of a half of a drop in the ocean that is the observable universe. How many other galaxies, and black holes, and Quasars, and stars, and moons, and planets are out there that we aren't aware of? Perspectives change when knowledge is obtained. That my dear friends, is insight.

{"Space exploration is a force of nature unto itself that no other force in society can rival."
– Neil DeGrasse Tyson}

The Dogon Tribe of West Mali

The Dogon Tribe origins are difficult to trace but what I have gathered of it is that they are most likely the Kemetic natives who fled the area after Kemet fell to invading forces, slavery was about, and to escape Islam's forced introduction. They mostly (at this current time) reside in an area referred to as the Bandiagara Escarpment, which is a sandstone cliff in what is referred to as Western Africa. The Dogon peoples are incredibly grounded in the principles of harmony (balance). It has been noted that "…in one of their most important rituals, the women praise the men, the men thank the women, the young express appreciation for the old, and the old recognize the contributions of the young."[20]

The Dogon people are so interesting because of how in tune they are with the celestial bodies, their movements, and how they were able to know this at such an early time in history before the invention of telescopes (and other modern tools utilized today to help see into

---

[20] Wikipedia – Dogon People. https://en.wikipedia.org/wiki/Dogon_people. Last modified on December 27, 2014. Accessed on January 17, 2015

[21] Dogon Village of Songo, with mud mosque, Mali.

the heavens). Because of some ignorant of respect humans, however, much of the information needed to trace these roots and knowledge facts are impossible if not highly difficult to investigate – it is essentially one of the biggest reasons that I am such a proponent of true and sincere respect amongst nations and their people.

The Dogon have accomplished retaining their sacred teachings and it is on us in the modern era to respectfully and sincerely seek out their truth and catalog it in the books for future researchers to take further. With that being said let us look at some of what they have been able to do.

"In 1976 Robert K. G. Temple wrote a book called The Sirius Mystery arguing that the Dogon's system reveals precise knowledge of the cosmological facts only known by the development of modern astronomy, since they appear to know, from Griaule and Dieterlen's account, that Sirius is part of a binary star system, whose second star, Sirius B, a white dwarf, is however completely invisible to the human eye…and that it takes 50 years to complete its orbit. The existence of Sirius B had only been inferred to exist through mathematical calculations undertaken by Fredrich Bessel in 1844."[22]

"In the late 1940's, Dogon priests greatly surprised the French anthropologists Griaule and Dieterlen by telling them secret Dogon myths about the star Sirius (8.6 light years from the earth). The priests said that Sirius had a companion star that was invisible to the human eye. They also stated that the star moved in a 50-year elliptical orbit around Sirius, that it was small and incredibly heavy, and that it rotated on its axis.

All these things happen to be true (the actual orbital figure is 50.04 +/- 0.09 years). But what makes this so remarkable is that the companion star of Sirius, called Sirius B, was first photographed in

---

[22] Wikipedia – Dogon People. https://en.wikipedia.org/wiki/Dogon_people. Last modified on December 27, 2014. Accessed on January 17, 2015

1970. While people began to suspect its existence around 1844, it was not seen through a telescope until 1862."[23]

It isn't at all coincidence that these people knew know this much about the universe and its heavenly bodies. It is not farfetched to accept this fact that they were right; that while some of their stories may currently seem untrue [*Look up: Nommo*] that perhaps it is our limited understanding and technology that prohibits us from seeing what they see. It is to this perception and this reality that we owe the debt of finding the answers for our modern era, the inner mechanics of the world auwa.

---

[23] www.sacredsites.com/africa/mali/dogon.html - Article that highlights the many aspects of Dogon life and Astrological practices.

[24] Rite of passage ceremonial site for Dogon boys becoming men, near village of Songo, Bandiagara.

Essence of Magnetism:

– The Unseen Atomic Micro-Universes –

Magnetism is the study of the "physical phenomena arising from the force between magnets, objects that produce fields that attract or repeal other objects."[25] Magnetism is an inspirational concept in the high sciences because magnetism affects every single type of material in some way. Its effect on certain types of material is so minute however that one might assume it is not affected at all (i.e. wood, rubber, plastic, etc.) and so in large part I believe that most people take the power and significance of magnetism for granted. For example, did you know that magnets are a basic component in disk drives? Did you know that once upon a time televisions had some powerful magnets within them used to create the images on the screen? Did you know that magnets stick to each other because they are exchange photons? What are photons? Photons are tiny particles of light – they are force carriers, meaning that they can exchange momentum or repelling force. Magnetics are so important because in large part earth's magnetic field is the sole aspect of defense that protects our atmosphere (and us) from the solar winds, and the intense solar effects of our life giver, the Sun.

On a more esoteric note a Latvian scientist, author, sculptor, and romantic by the name of Edward Leedskalnin created what we now call the Coral Castle, currently located in Homestead, Florida. The actual magic behind this magnificent place is how one man that was around only 5' and 100lbs could single-handedly (with common small tools) lift and move into perfect place pieces of coral and rock that were sometimes upwards of 25 tons!

---

[25] Lewis, Tanya. "What is Magnetism?" LiveScience. July 9, 2013. Accessed January 22, 2015.

{"I have discovered the secrets of the pyramids, and have found out how the Egyptians and the ancient builders in Peru, Yucatan, and Asia, with only primitive tools, raised and set in place blocks of stone weighing many tons!"

– Edward Leedskalnin}

Ed wasn't the only one that seemingly cracked one of the universes most closely held secrets (reserved for only those who dedicate many sleepless nights and lost friends) – no – Nikola Tesla was another illustrious genius who was able to create such magnificence in his lab.

Tesla, who was born on the 10th of July 1856, was a Serbian-American who is noted as an esteemed futurist, mechanical and electrician engineer, and visionary inventor. He is the father of the Alternating Current which is utilized in almost every electronic device in our modern world and not only did he invent that but he also foresaw the ability to send pictures through signals at a time where people often called him crazy for his ideas of what is truly possible when one is enlightened in the scientific arts. One of my favorite Tesla inventions was what is known as the Wardenclyffe Tower (slang: Tesla Tower) and was believed to be able to send electricity wirelessly across the globe enabling all of earth's inhabitants to have access to clean wireless electricity. This was a true humanitarian who was defunded and in constant battle with the greed of his financiers and Thomas Edison who attempted to devalue Tesla's Alternating Current by electrocuting an elephant to death. Tesla believed so grandly in the science of magnetics and based much of his work around it.

{"Ere long intelligence-transmitted without wires-will throb through the earth like a pulse through a living organism. The wonder is that, with the present state of knowledge and the experiences gained, no attempt is being made to disturb the electrostatic or magnetic condition of the earth, and transmit, if nothing else, intelligence."

– Nikola Tesla}

Tesla often taught far over the heads of the people he was speaking to and so he came off arrogant to many of those same people but post mortem we are able to begin innerstanding his true brilliance and visionary ideas. Some of his sayings still haven't quite been decoded yet, and it is our responsibility to do exactly that. One of my favorite quotes by Nikola Tesla:

{"If you only knew the magnificence of the 3, 6, and 9, then you would have a key to the universe."

– Nikola Tesla}

Magnetism involves looking at the most minute instances of magnetism and the most massive of the same. The human heart, for example, has a magnetic field that can be measured feet away from the body. Another example is our own home planet, Earth. It generates its magnetic field from the Earth's innermost core grinding against its molten core, which creates a magnetic field; that same magnetic field is what protects us from the Sun (our life giver) that also has the power to strip our planet of its atmosphere which we all know would certainly be our end. Even our brains have a magnetic field that can be measured by medical equipment and in all of these examples one must wonder what is the purpose of it? Why does this happen? What does it mean? Can it be manipulated? Enhanced? What happens if it turns off? What is Earth's core made out of in order to generate this field and why are two of them grinding on each other? Are there magnetics in space? All of these are perfect questions, and we need to do more to find out the answers. Magnetics are, after all, everywhere – learn how to innerstand magnetics and you'll become… different. Perspectives change when knowledge is obtained; a phrase I came up with and began saying to myself before I begin any new project and now it is time that I reveal to you the second portion of that phrase. I didn't create this one but the second part is "Perception is Reality". What you perceive is what is projected, it is what is real. I look at a cancer patient that perceives their reality being that cancer is about to lose their fight – much research is being done that aims to innerstand how the human body is able to fight off such diseases with the mere thoughts of winning. All of that to say, change your perception, and you'll thence change your reality.

{"Change your perception and you'll thence change your reality."

– Heh Heru}

{"There is a magnet in your heart that will attract true friends. That magnet is unselfishness, thinking of others first; when you learn to live for others, they will live for you."

– Paramahansa Yogananda}

{"I believe that there is a subtle magnetism in Nature, which, if we unconsciously yield to it, will direct us aright."

– Henry David Thoreau}

{"I don't care that they stole my idea, I care that they don't have any of their own"

– Nikola Tesla}

Earth Magnetics

The Earth has a magnetosphere that is responsible for protecting us from the severe damage the Sun would do to us. The motion of the core and type of material that comprises our Earth's core generates our Earth's magnetosphere. After our magnetosphere was formed volcanoes and other events expelled the gasses into the newly protected area that would become our atmosphere; allowing the formation of liquid water.

"Earth's Magnetosphere : Image of the Day." Earth's Magnetosphere : Image of the Day. Accessed August 13, 2015. http://earthobservatory.nasa.gov/IOTD/view.php?id=50208.

The Sun is constantly bombarding our planet with deadly radiation and it is quite literally the power of the magnetosphere that keeps us safe from certain death in that regard. Without that shielding, the solar wind would obliterate our atmosphere, which would cause the liquid water on our planet to boil off and life, as we know it would end not soon after.

Steele Hill/NASA

Our thick atmosphere creates the beautiful pressure bubble that helps to keep our liquid water in that liquid state. It allows for rivers to flow, oceans to exist, and rain to fall. If the magnetosphere was ever eliminated the solar wind would blow this thick atmosphere away. Not having the atmosphere would remove the pressurization needed to keep our water in the liquid state, and would overall turn our planet into what Mars has become today; a seemingly [*I say "seemingly" because in the matters of the universe we are infantile in our awareness. We cannot see much of the universe. We cannot travel so far outward into her grandeur, and we don't really seem to have a true intention to want to explore space all that much aside for defense and war purpose*] desolate planet that can barely sustain our current form of human life.

We can see on this scale the power of magnetism. It is among one of the many components that have to be in place in order for us to exist in our physical form. Since it is of such great importance, we should study it with true heart. We should find out what it is that our ancient ancestors knew about this power. Imagine if numbers were never created, or imagine if computer technology was never created, or imagine if the concept of time was never created; could you imagine how different life would be? We cannot protect, preserve, and revere something we don't understand and that is the reason that we must study, protect, and revere the magnetosphere.

I would like to reiterate how important applied knowledge is with a bit of history. Let's time travel back to July of 1904. A publication by Sherwin-Williams wrote of the immense dangers of paints containing lead. Before 1904 leaded household paint, gasoline, and children's toys was the normal. People didn't know how dangerous lead was until a team of scientists with true heart made them privy of it but by that time so much damage had been done. So many people became physically challenged, or succumbed to death.

When you know better, you do better.

We, in everything that we venture into, must not succumb to the distractions of the matrix that we reside in. We must focus on the high sciences that we create a better condition for ourselves, our people, and for whatever other life essences exists auwa. We owe it to our creator to be good masters of applied knowledge. That's insight.

Local Magnetism

Local magnetism is comprised of magnetic fields independent of the Earth's magnetic field or the magnetic fields occurring in outer space. Local magnetism is comprised of things such as the human heart's magnetic field, our brains magnetic field, and other micromagnetic occurrences that exist around us. These local occurrences of magnetic fields affect us in such profound ways that modern scientists, healers, and medical doctors have invested great time and energy into better understanding the how's and why's of it all. We've found out some interesting factoids about them thus far as well, and I want to take a minute to share with you what those factoids are as well as why we should since we all have hearts and brains should seek out this knowledge of self.

We commonly think of the heart as the organ that just pulses at a precise schedule to pass around the oxygenated and nutrient-enriched blood to all of our other organs. However, we are now learning how limited this innerstanding of our heart actually is. The ancients seemed to have already known about the true power of the heart beyond its ability to pump blood from place to place within us. Modern scientists, healers, and medical doctors are now beginning to innerstand that the heart in many ways is able to take charge as it relates to many functions within our body. The question, in my opinion then becomes who or what controls how we function? What I mean by that question is what tells the heart how to function, what tells the brain how to function, and what directs us in our "conscious" mind? Asking these types of questions makes us stronger in our innerstanding of just what is happening auwa. Existing on that frontline of our human ignorance and chipping away at those borders opens up new worlds for us to explore and that my dear friends is insight.

The First, the Heart.

When a human embryo is forming within the mighty womb of the mother, the very first organ that is created is the heart. When the embryo is but a few cells in size it can pass the essential nutrients to all needed locations quite efficiently but as the embryo grows so does the need to pass those nutrients around more efficiently; enter the mighty heart. The heart becomes responsible for transferring the nutrients that are essential to the health of the embryo. The heart, through its normal operation also emits a powerful magnetic field that is full of information that is measured by an electrocardiogram or ECG for short.

Rollin McCraty wrote in the paper, *The Energetic Heart: Bioelectromagnetic Communication Within and Between People* that, "The heart is a sensory organ and acts as a sophisticated information encoding and processing center that enables it to learn, remember, and make independent functional decisions." She later explains that "The heart, like the brain, generates a powerful electromagnetic field… The heart generates the largest electromagnetic field in the body. The electrical field as measured in an electrocardiogram (ECG) is about 60 times greater in amplitude than the brain waves recorded in an electroencephalogram (EEG)."

There is still more exploration and research of just what this information is comprised of and what its purpose is, but we know for a fact that nothing the body does is for no reason.

Copper Pipe & A Magnet Experiment

Something strange, magnificent, and sublime happens when you take a simple copper pipe and a magnet and cause them to be in a particular form. As a baseline, if one were to measure the length of the copper pipe and drop the magnet outside of it (in that same distance) it drops at a fast rate, a normal rate according to gravitational rules. Now, a copper pipe isn't magnetized like many other metals, but when you drop the magnet inside of the copper pipe, it falls at a rate much slower than if it were outside of the copper pipe. One could even watch the magnet and observe what happens when it is inside of the copper pipe; it wobbles and seemingly breaks the rules of gravity. What's happening in this experiment has many names with the modern names being Lenz's Law or Eddy Currents and the more ancient names being (for the moment) elusive as they have been lost to the sands of time. There was a really great article by education.com, and I want you all to take a minute and follow this so that you receive the observational science perspective. Now, if you cannot perform the experiment, then that is alright as you can simply utilize YouTube it and find a plethora of people showcasing exactly what happens when you perform this experiment.

From education.com:

It is 1834, and you have just heard of this marvelous new phenomenon called eddy currents. Some fellow named Lenz discovered them, and you're curious if you can find out something special about them yourself. The friend who mentioned it to you said that some interesting things happened when magnets and copper pipes interacted, which is strange, because copper isn't magnetic! It's time to find out what's going on.

Materials

- Copper pipe or fresh tube of aluminum foil
- Aluminum Cookie Sheet
- Aluminum or non-magnetic metal washer
- String
- Small Neodymium Magnet
- Voltmeter (optional)

Procedure

1. Touch your magnet to your various materials to confirm that they're not magnetic.
2. Take your magnet and drop it from about the height of your pipe onto something soft.
3. Now, drop it through your pipe or tube of aluminum foil. *What do you expect to happen? What actually happens?*
4. Place your magnet on the cookie sheet and tilt the sheet so that it slides off. *Does it behave oddly? Why do you think that is?*
5. Tie your magnet to the string, and swing it in a low circle so that it passes over a metal washer placed on a smooth surface. *What happens to the Washer? Can you think of how this might be related to the other behavior you observed earlier?*

Results

When you drop your magnet through a copper tube, it slows down. The magnet will also slide down the cookie sheet slowly, and nudge the metal washer in the direction the magnet is spinning. The voltage will spike when the magnet is moving next to the metal, but not when the magnet is sitting still.

Why?

Magnetic fields are the result of electric currents. Changing a magnetic field (moving a magnet) next to a non-magnetic metal will induce an electric field (a voltage difference) in the metal, which subsequently generates a magnetic field with an opposite orientation with respect to your magnet.

When your magnet moves next to a metal it creates these fields, but the fields act in a very specific way. They want to cancel out the magnetic field in the metal, because metals don't like having electric or magnetic fields inside of them (that's why electricity flows through metals easily—they're trying to cancel out the difference in electric potential by moving electrons around!). This phenomenon is known as Lenz's Law.

The magnetic field induced in the metal attracts the falling magnet, creating resistance. This resistance is what slowed down your magnet. As your magnet slows down, it stops generating as much current, which reduces the resistance acting on the magnet's movement. Gravity speeds the magnet back up again until it reaches a happy medium speed. Basically, your magnet is creating a whirlpool of electrons around it as it falls through your pipe. Neat, huh?[26]

---

[26] Jacobsen, Alex. "Lenz's Law: Magnet Through a Copper Tube." Lenz's Law: Magnet Through a Copper Tube. Accessed February 3, 2015.

Essence of Music & Audiology:

– The Frequencies that Move Matter –

Music is simply defined as sound organized in time. Audiology is the science that focuses on hearing. Humanity was able to discover this knowledge (although it is safe to assume that it isn't new knowledge so we really didn't discover anything but rather came to innerstand it – that which was already existing within nature) and thus we began to manipulate it at our will.

We created various musical instruments able to manipulate the elements, and we created scientific instruments able to assist us in our human audiology endeavors. However, it is my opinion based upon my long travels that we also came to a fork in the road as it relates to the direction or morality of these realms. I refer to these directions or moralities as the matrixes or the Taijitu. That fork in the road comes in the types of manipulations that we allow to occur.

Visualize it this way; you discover a road off of the highway and about a mile into this road it comes to a left path and a right path. One sign reads "Positive Progress," the second reads "Negative Nancy", and the U-turn sign reads "Neutral Dude." Some people take the left, some people take the right, and then some people make a U-turn and leave. These are, in my opinion, the three matrixes that people inevitably find themselves in as it relates to music and audiology; you're either in it for the good, the bad, or you just don't care to notice either.

Within the musical spectrum, we have what I see as the Taijitu of two paths, one being of positive energy and the other being of negative energy. I submit that positive and negative are relative to our current perspectives and perception and based in the human form, but I ask that you walk with me on this.

Positive musical energies are things like binaural beats, healing tones, Cymatics, meditation chants, and the like while negative musical energies are comprised of any musical element that causes one to resolve in fear, unjust hatred, disharmonic Hz ranges, or generally the opposite of love.

Within the audiology spectrum, we have the positive and negative actors as well. The positives being any action, invention, or field of study (e.g. Cymatics [*by the way e.g. means "Exempli Gratia," which is Latin for "For Example"*]) that promotes the well-being of the Earth and her inhabitants and gives rise to knowledge uncorrupted. The negative energies being any action or invention that creates decimation of the Earth and all of her inhabitants or the stagnation of acquired knowledge.

An example of the positive energies in both fields can be seen in Cymatics, the study of how sound manipulates matter. This field of study has caused us to innerstand a little more what affects sound waves, and vibrations have on the matter that formulates the world around us and within us. Through this field of study, we began to wonder did "sound" have an integral part in creating life since frequencies have the unique ability to mold matter into patterns and hold them in order [*I will cover this in greater deal within the Cymatics subsection*].

An example of the negative energies in music would be the radio tune where Artist X sings about how many women he/she has (referring to them in a derogatory manner), or how they would love to "tap that ass at the club", or spews insensitive lyrics. An example of the negative energies in audiology would be the Sound Cannon, a "crowd control" product used in a way that bombards the subject with such an intense level of sound that it deters them from any further action. Another example being the "Flash Bang Grenade" which is utilized to overpower the sensory organs of another sentient being thereby giving an upper hand to one life over another. Again, positive and negative can be seen as relative depending on the moral compass or situation of the actor. However, in this peace centered sense I tend to think of anything that causes harm upon sentient life as negative.

Both of these examples create no moral good, and since they create no moral good they are thence negative.

The question is where do you reside? To which side of energy do you belong to? Are you on the side creating positive musical aspects and progressing scientific (and spiritual) innerstanding or are you on the side of creating music of no moral value, and audiological products whose sole aim is to cause harm or discontent?

In studying the sublime art high science that is music or audiology, it is critically important to first innerstand their history and where they currently stand now. Specifically, what role music had to the ancients as compared to its role in society now and what role audiology had in the past as compared to its current application. This means that we must study its history broadly in order to form our own perspectives on the matter; perspectives based upon source material and reason. This is what I have done, and this has led me to the perspective that you have just read in the introduction above.

In this section of the book I will cover some high science concepts that I've stumbled upon that caught my interest for two reasons: One being that on the scale of a global populous they seem to not garnish much attention or research and the other being because I am an avid practicing musician that innerstands the power of controlling one's mood or actions by the way of musical audiology.

Think of how happy you get when you hear your favorite song from the past on your music player. Think of how sad you get listening to a break-up song that reminds you of a past love. Think of how inspired you get when listening to an epic soundtrack to a movie like *Inception*. The fact that the musician or arranger was able to tap into your being and guide your emotions through pitches and tones organized in time is deserving of study and reverence. I feel that we have slightly lost our way and take the science of it for granted. It is for these reasons that I have crafted this section of the book. Enjoy.

432Hz & 440Hz

432Hz is known by many of the initiated as the frequency that is mathematically tuned in with the universe and essential for enlightenment and spiritual wakefulness. There are many articles and aspects of research being performed all over the world that showcase this to be fact.

In Ancient Kemet, Greece, Tibet, and many other societies the standard tuning was 432Hz, and it was utilized for more than just musical entertainment. Many of their musical instruments recently unearthed are harmonic with this standard as well. Many of these societies utilized music in this tuning arrangement for healing practices and higher spiritual purposes. There has been much research done on the subject and even more needs to be done alongside with field research to discover why it was 432Hz [*I have found evidence to showcase that instruments of those societies were tuned to 432Hz*], and why J.C. Deagan changed it to 440Hz and got the rest of the world to so easily just go with it.

If you aren't familiar with who J.C. Deagan was, he was musician and manufacturer that was born in Hector, NY on November the 6th, 1853. While on active duty with the Navy he attended a lecture by a German physicist and physiologist called Hermann Helmholtz. Helmholtz was the author of "The Theory of the Sensations of Tone as a Foundation of Music Theory" and J.C. Deagan was compelled to create bells and chimes in the 400Hz tuning and set out to make this the new standard. He married into a wealthy and powerful family that enabled him to persuade various music organizations to change the tuning pitch from 432Hz, where it had been for thousands of years, to 440Hz.

I haven't found any substantive reasoning behind his drive to change the mathematically and scientifically sound 432Hz tuning standard to the irregular 440Hz; a move completely out of sync with history and master musicians of the day.

I have however come across a letter address to the Italian government from Giuseppe Verdi. He writes:

"Since France has adopted a standard pitch, I advised that the example should also be followed by us; and I formally requested that the orchestras of various cities of Italy, among them that of the Scala [Milan], to lower the tuning fork to conform to the standard French one. If the musical commission instituted by our government believes for mathematical exigencies, that we should reduce the 435 vibrations of French tuning fork to 432, the difference is so small, almost imperceptible to the ear, that I associate myself most willingly with this."

His most noble effort, however, fell upon deaf ears and in 1917 the American Federation of Musicians accepted the A440 as their standard. Sometime around the 1940's the United States showcased this new standard across the globe and it eventually became the new world standard in 1953. But why is the question? Why was there ever a need to shift away from a tuning standard cast forth by thousands of years of musical study and reverence to the high sciences? Especially given that there was no reason or notable explanation for the change. Napoléon Bonaparte once stated that history is written by the victor and that it is a lie agreed upon; I wonder if this change has an ulterior motive. I don't profess to be an expert on this matter but as a musician, child of the light, and student of the sublime arts I am admittedly curious and skeptical of why such a change was necessary without public debate. We should find out together, and I implore you to join me.

Frequencies & Vibrations Introduction

I thought it important that you realize how important frequencies and vibrations are to the world within us and around us. Sound waves quite literally have the ability to manipulate matter into various forms and patterns. They seem to also be able to hold aspects of matter in formation. There is an excellent music video by Nigel Stanford called "Cymatics" that I implore you to watch as it showcases the beautiful relationship that sound and matter have with one another. NASA developed another great example of this, and it's called "Planetary Sounds." It sounds strange at first glance but our universe, in fact, makes tons of noise and perhaps that noise plays a far bigger part in the creation of that which is auwa.

While it is safe to say that the ancients knew about frequencies and vibrations much of that knowledge was lost, stolen, or is still hidden by actors all over the world. Whether it was by war, conquest, thievery, or corruption truth and light always reveal themselves to those who can see their faintness among the corruption.

The modern father of Cymatics, Hans Jenny once said:

"The film you are about to see has no characters, it has no people. It is a film to describe to you, and explain visually, the effect of Cymatics frequencies and texture, structure, water, and oil. If you spare a little of your imagination as you watch this film as it runs, you will see many things that answer many questions. You will see living forms, living amoeba, almost animal-like creatures. You will see continents being formed, The Earth itself coming into existence: explosions, eruptions, atomic explosions, and bombs. You can see all this and watch it before your eyes. Everything owes its existence solely and completely to sound. Sound is a factor which holds it together; sound is the basis for form and shape. 'In the beginning there was the word, and the word was God': We are told this is how

the world began and how creation took shape. If we put that into the modern idiom and say that into the great voids of space came a sound, and the matter took shape."

You simply just have to listen, and you will hear the immense noise of the universe in motion.

Remember that our human hearing range is rather limited. For a person that has experienced no hearing loss, the range is 20 to 20,000 hertz. There are sounds far beyond our immediate hearing range and it is safe to assume that they are playing an active role in the universe. The questions of what is exactly that is, to what effect, and what role they may have played in our creation are questions that I would love to know; thence, I continue the search and so should you.

Cymatics

The word Cymatics derives from the Greek word meaning "wave" and it is the realm of research that focuses on the role frequencies, and vibrations play in the manipulation of cosmic elements. That is to say the study of how sound moves matter around or holds it in place. The term came into normality by the brilliant physician and natural scientist Hans Jenny. He was inspired by the works of Ernst Chladni, a German physicist, and musician who specialized in the auditory arts. Ernst Chladni is often referred to as the father of acoustics because of his immense work on vibrating plates and the properties of sound in motion. It was because of him that the modern day study of sounds manipulation of the elements is available.

Take a second and look at the pictures of what Cymatic frequencies do to matter on the following few pages. I advise you to watch Nigel Stanford's music video called Cymatics and the documentary called, Cymatics: Bringing Matter to Life with Sound. I also encourage you to look up more on Cymatic frequencies on Google or YouTube as since we live in the age of mass consciousness where information on virtually any subject can be easily retrieved in a matter of Nano-seconds.

Ernst Chladni   *Hans Jenny*

Chladni Plate by: Matt Venn (Creative Commons BY-SA 2.0)

[*In the photo above sound waves were played through a medium with sand upon the surface. The sand reacts in alignment with the mediums vibrations caused by the sound and you thence receive the visual product*]

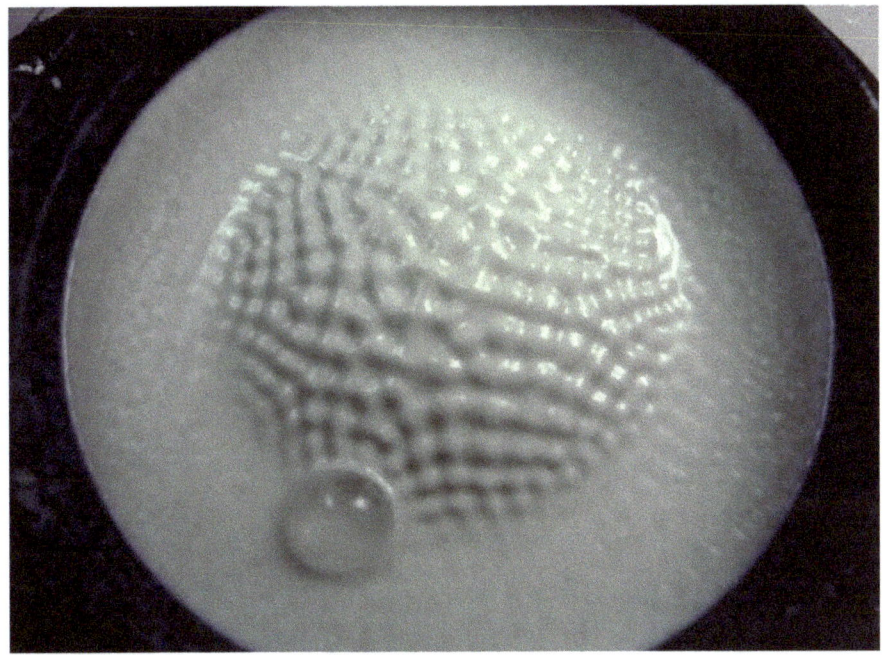

Author: Collin Cunningham (CollinMel)

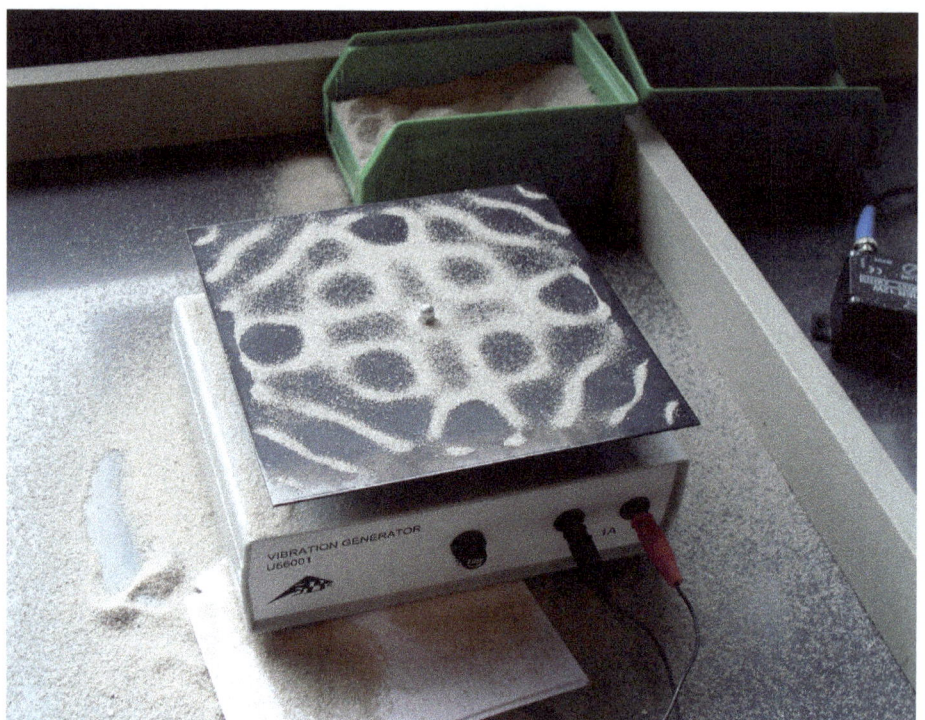

Public Domain: Retrieved 2015

Public Domain: Retrieved 2015

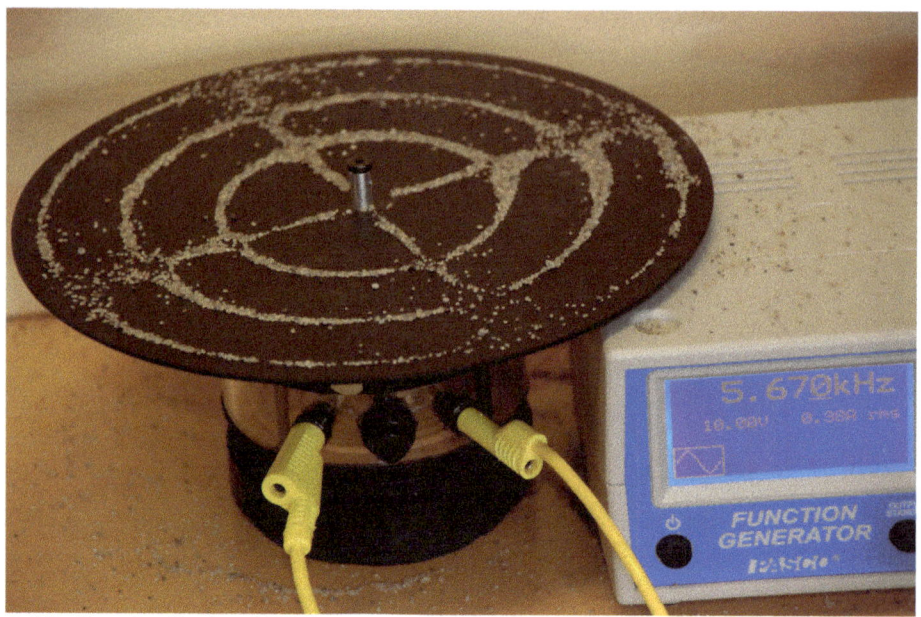

Public Domain: Retrieved 2015

In studying this specific intersection of music and audiology, we begin to innerstand what role both music and sound have in the manipulation of cosmic elements. Seeing as to how we Homo-sapiens are also made of the cosmic elements, investigating the relationship between sound and matter deserves reverence and due attention. Sound is an integral part of our human experience, and even our bodies know this on a sub-conscious level.

"The vestibule is the region of the inner ear where the semicircular canals converge, close to the cochlea (the hearing organ). The vestibular system works with the visual system to keep objects in focus when the head is moving…"[27]

Our bodies have developed complicated organs and systems in order to help us maintain our balance and perception – I'm speaking

---

[27] National Dizzy & Balance Center. Accessed October 16, 2015. http://nationaldizzyandbalancecenter.com/resources/balance-system/.

specifically of the ears here. I use this example because it shows how much we take the analysis of sound for granted; it shows how much we generally ignore the sublime nature of sound and its relationship with the operative auwa. There are many more examples of where sound plays an integral role in the nature of things, and I implore you to find them, and share them!

I implore you to research this science more and do reach out to me with good conversation, questions, and answers; this is how we grow.

Questions on Consciousness

– My Perspectives; Your Questions –

What exactly does it mean to have a conscious? To be conscious? What is consciousness? Is being conscious the same as being aware? If so, then what exactly are we conscious of or aware of? What responsibility do I have if I am found to be conscious or aware? Why are we here? Where did live originate? What is really going on here? Can you tell us more about Horemaket? Can you explain some cool facts about the Great Temple at Giza? Should I go Vegan?

While speaking with friends, strangers, and at many various events these are some of the questions that I received that come about so frequently I thought it important to discuss it within this book as "consciousness" or "wakefulness" or "awareness" are all inspirational concepts in the high sciences and they thus deserve analysis and study.

Q1: What exactly does it mean to have a conscious?

A1: To have a conscious is to be morally upstanding [*as it relates to the rejection of causing harm upon sentient beings*]. To have a conscious is to practice Dharma through innerstanding and not by fear of consequence or some other punitive motivation. To have a conscious is to have empathy, respect, and a prime innerstanding of what it is to be "Good". Avoiding negative vibrations and lower vibrations, which are signature to fear, hate, evil, and all things not in the good is how one retains a morally upstanding conscious. To have a conscious is to care about life, matter, and to see/feel yourself in all else. This is what it means to have a conscious to me, and I invite you to build with me on this concept.

Q2: What exactly does it mean to be conscious?

A2: To be conscious is essentially to have a conscious with the addition of being in tune with a more intense set of knowledge. What does this mean? This means that one is aware that all matter has a common root (as it relates to the building blocks of life and to humanity's known origin as well). Also in that intense set of knowledge tool box (ISKTB) is the awareness that one has obtained on how important the seven Masonic liberal arts are, how important it is to remain a true Gnostic, a true steward, a truly sincere lover of the inspirational concepts in the high sciences. To be conscious is to have an ISKTB that includes a diligence and lasting thirst for more knowledge and information – with the addition of actually living in the knowledge attained and not being a hypocrite to its teachings and customs.

Q3: What is consciousness?

A3: Consciousness is wakefulness which is awareness which is generally intelligence married to wisdom. Wakefulness as a note can mean that one has found euphoria in knowledge and not only feels better in life but also has left a level of consciousness that they once resided on/in. Think of it this way: If Amenhotep decided to venture outside of the only city he has ever known and walks into another city and another city and another city they have achieved a wakefulness about them that has enabled them to realize that there's more to it all. They become awake. Awareness is the same as wakefulness essentially. Amenhotep became aware that there is more than just his city when he ventured outward into the world and began to experience all that life has to offer outside of his previous matrix. Consciousness is the beautiful marriage of wakefulness, awareness, and intelligence married to wisdom. It's a lifestyle that has a strict set of unwritten rules and regulations that one must abide by. Rules not meant to govern the individual but rather meant to guard the integrity of what it is to be in consciousness.

Q4: What are my responsibilities as a sentient being?

A4: To not be evil, to not reside in the realm of willingly ignorance, to not be motivated by illogical bias, and to be guided by reason and not desire. That is to say, [*as we philosophers say*] follow the common good. Your responsibilities also include to thwart evil actions within yourself and within the world around you; not guided by silly bias or manipulated feelings based upon evil objectives. You will never be able to kill evil off completely as evil and good are in a perpetual dance with one another – the objective is to innerstand this but to seek out the proper dharma through that innerstanding. Live in the magnificence that is peace and positivity – not in the depths of despair that is negativity and evil; at least that's my view on it.

Q5: What are some cool facts about the Great Temple at Giza?

A5: There are so many incredible things about that structure that I can't spend too much time here writing about, but some of the most noticeable ones that I've come across that deserve more research, validation, and study are as follows:

1. It was constructed around 2589 – 2504 before the mythical (or literal, whichever you choose to believe) Christ and it is the only great wonder of the ancient world still in remarkably flawless condition despite the fact that the Arab Muslims chipped away the shiny limestone casing and used it to build their mosque and temples. That's pretty neat.

2. The Great Temples of Menkaure, Khafre, and Khufu are all magnificently and perfectly aligned with the Orion constellation.

3. The binding mortar that the ancient Africans used to keep the stones in place has been researched and their components broken down but the actual mortar still to this day cannot be recreated.

4. It's aligned with in 3/60th of a degree of error with true north. A design feat we to this very day cannot come close with.

5. It was built precisely in the center of Earth's landmass.

6. Various measurements calculated throughout the temple showcase that the builders were fully aware of pi and the Golden Mean – bear in mind that this was thousands of years before Pythagoras was ever born.

7. The builders knew that the Earth wasn't flat! They knew that it was in the shape of a sphere and the temple was constructed in such a way that it allowed them to chart and measure astrological events.

8. The four faces of the temple slightly slope inward making it an 8 sided temple. This feature makes it even that much harder to build and it was the only temple found-to-date that was constructed in this way.

It's important to note that we cannot recreate that temple today. We have no idea how the ancient Africans were able to do so much and know so much about the mechanics of the universe. If it wasn't for the Persians, Greeks, Arabs, Romans, and other invaders demolishing their civilization to the point of ruins and desolation we might have still had the original builders to call upon. To me, it's quite sad how impossible it seems for humans to just quit trying to kill or one up one another. It is by far the saddest and most pitiful pissing contest I have ever been forced to witness.

Q6: Should I go Vegan?

A6: I would say so yes; especially given how cruel the world treats its animal friends. I tend to look at things from a naturalist perspective but also from a spiritual plane; in that your personal choices whether you are aware of them or not have consequences in the physical and spiritual realm. Life is a spiritual thing and in consuming the dead flesh of a sentient life form you inherently consume the negative energies surrounding them.

I find it odd that we humans have no nutritional requirement to consume meat, and yet we do. I found that humans eat meat dairy and seafood out of desire and not necessity. Our teeth aren't even designed for tearing apart flesh – rather – they are geared towards mashing, which is signature to mashing plants or plant derived material. Furthermore, we currently as of the year 2015 raise more than 70 billion land animals, and we take from the ocean over 2 trillion pieces of sea animal life just for our ever growing appetite; keeping in mind that this is from desire and not necessity. What this practice does is creates ocean acidification, the release of the extremely dangerous nitrous oxide into the atmosphere, completely demolishes fresh water supply, requires massive amounts of space [which is why so much of the Amazon rainforest has been demolished], and the animals need so much in food supply just to grow. How is any of this sustainable and how can we call ourselves caring beings if we participate in such ridiculous actions?! I won't get into the particulars of what happens in slaughterhouses across the world, but the evidence is there in the form of videos, books, articles, and so much more if you simply do a Google search for it.

Amissio Status Quo

– Rejection of the Status Quo –

The following section of this book is a collection of excerpts, quotes, articles, and general thoughts from myself that venture into a mindset that rejects the status quo and wishes to inspire into you the same perspective. They are thought experiences in essence that you should truly internalize and analyze within yourself while sitting outside in nature; preferably in the woods or a field, or something generally with an awesome view with the temperature being around 68-77 degrees Fahrenheit. The Sun or overcast should be present, and you should be by yourself with a solid premium pair of headphones on; preferably the Sennheiser HD498's, the Beats by Dre Studios, or the equivalent Bose, Sony, or Shure over the ear with 40mm drivers' brands. The music that should be playing should be an audio vibration that I have created specifically to enhance your chakratic [*chakratic – another made up word that I just created that means to allude to the chakras or meditation on a higher level in essence*] awareness. The audio vibration is called "Dimensional Drift Audio Tone" and by purchasing this book you are entitled to download it for free (request it via the 'Contact Us' page of saamr.org). The entire time that you read this section of the book have this audio tune playing and also play it whenever else you wish to calm down or find the heavens; whether your sleeping, driving, stuck in traffic, stressed out, or need to focus I have designed this audio vibration to match many different situations. In addition, there are two albums which can be found on iTunes by Nigel Stanford, one is called Solar Echoes and the other TimeScapes which I also recommend as they have been integral in assisting me while I compose this book. Probably the most played music while composing this book were the instrumentals by the soulful producer J. Dilla [*songs: 1. Purple (instrumental), 2. Life (instrumental), 3. Flowers (instrumental), 4. Colors of You (instrumental), 5. Runnin' (instrumental), 6. Smooth (Unreleased), and 7. Sunbeams (instrumental)*]. Not only those but also Chillhop artists such as Vanilla, Joseph Jacobs, Nymano, and others were on constant rotation. Truth be told, they've also inspired me into daydreaming episodes during the day or while I meditate and I implore you to check them out as well as discover your own positive vibration music.

After you complete this section of the book, I want you to take a walk with your favorite tunes playing or just take a tour bus ride somewhere and just take it all in. Look around, look at the sky, look at the dirt, look at the advertisements, look at the people, and take it all in; finish it all by looking at yourself naked in your mirror. As you overlook yourself move closer to the mirror and make the pupils of your eyes focus on their reflection and think about who you are – what you are.

When you are ready, turn the page and begin with me a journey into the Amissio Status Quo. – Love, Heh Heru

Thy Black Mirror

I live my life looking into my black mirror…
There's no reflection of myself as in the vein human form though…
You know – no clothes, no bills, no distractions, or ignorance…
Mine eyes see nothing but the black mirror…
Carbon and so-called empty space…
My mind is aware that nothing physical is truly real; perceptions.

- **Cedric Paul Harriott**

This is a chant that I perform while in meditation. It allows me to remove myself from the bound state that is the human form. I reside in the realm of thought that the older we get, the more we transfer energies into the plane of the sprits [*plane of the spirits being wherever our life essence originates*] and that through proper meditation we don't have to wait until we age out of the physical form, but rather we can tap in and out of that sublime area.

What do I mean by "Thy Black Mirror"? Thy means Mine Own. Black, signifies space, a void of matter and yet all of matter; carbon. Black, it signifies life and yet it signifies death, dead and all of that which we don't yet innerstand; the universe – you know – the heavens above us in the night sky that we see when we look at stars, planets, nebula, and all else that we can't fly out and study in our physical form. In my chant it allows me to visualize the concept of Zero that I spoke on in the Essence of Numerology.

A mirror is an object that we utilize to evaluate the look of ourselves and in the pensive thought of evaluating Zero, I evaluate myself. When I look out into Thy (mine own) Black (Life, Death, Transcendence) I see myself and all else around me and within me, I see dimensions, EYE see dimensions of mine own life, death, and transcendence in bounds with all else around me and I am thus granted the ability to innerstand that which is around me and within me. A journey I still partake in every day. This is what Thy Black Mirror means.

What do I expect from you, my reader? I expect you to do and feel the same. This mirror is your mirror as well and in order to access it all you have to do is be aware and watch, as it will manifest itself first in your mind, then in how you live your life. I expect you to not just read this and forget it to the annals of memory and time in the future nay, I expect you (and you should expect yourself) to live this. I expect you to mentally travel to a sensory deprivation chamber and experience high science thought and its ability to create episodes of lasting self-exploration.

The Illusions of Perceived Realities

If 99.99% of all atoms are comprised of so-called (SC) empty space,
THEN
Aren't I, being comprised of atoms, 99.99% SC empty space?
THEN
Isn't the chair that I go to sit on right now 99.99% SC empty space?
THEN
Why don't I fall straight through it? Why can't I run through walls?
What is holding it all together?
Where does my life force come from?
Where does it go when I leave my physical form?
What, pray tell is going on here!?

- Heh Heru

First of all, mind = blown, or at least one would assume that but in actuality it can be explained quite simply at an atomic level [*although it is still mind numbingly sublime how intense of a concept I am about to reveal to you, thy reader*].

You'd better be sitting in the grass outside, with a beautiful view around you on a nice weather day before you read this next sentence and if you aren't then be a good reader and stop reading at this very moment until you are.

Before, I reveal to why we don't simply fall through forms of perceived matter I want to plant a foundation. An Electron is a negatively charged particle that surrounds an atoms nucleus. A Nucleus is the central most part of an atom – It contains most of an atoms mass, and an atoms nucleus is also, composed of protons and neutrons. A Proton is a positively charged particle found within an atoms Nucleus and a Neutron is an uncharged particle found within an atoms Nucleus. What ingredients create these atoms? Gluons and Quarks are what make Protons and Neutrons but modern day science isn't certain what Gluons or Quarks are composed of. We don't know if they are energy, the keys to multi-dimensional space [look up Superstring Theory later], or much else on them, which is why we need you, the reader to ask questions, and join the journey to research what exactly is happening here. Why is it important? Great question – Consider this you, myself, the tree, the dirt, the magma, the air, the birds, the rhinos, the dinosaurs, light, I mean only EVERYTHING is comprised of this same basic material that we don't have any true handle on innerstanding. Shouldn't we explore to know everything about what is around us and within us since we have been endowed with our massive brains that hold more synaptic connections than stars in the universe (well at least the observable universe that is)?

Okay, now we I'm ready to tell you why we don't fall through chairs or can't walk through walls!

We don't fall through the chair we sit down on, and we cannot run through walls because we are met by the awesome resistance of force fields in constant activity. That is to say that all things are literally levitating on a lasting electrostatic field that don't actually touch. For example, when you grab a doorknob to open a door you really aren't even touching it! When you sit on the floor that feels so solid, you really aren't touching it! When you go to drive your car and grab the steering wheel you really aren't touching it! A shell of electrons surrounds every atom, and this cloud of electrons creates a negative charge outward into the world. It is fact that particles that have the same charge repel each other [think, magnets that repel each other, no matter what] and when two atoms attempt to come upon each other their electron clouds repel each other. This leaves us with a world of matter, which we think we touch but we really don't – it's a beautiful illusion of perceived realities that alludes to how perspectives are being changed when more knowledge is attained.

Sun

I look at the Sun and my mind flutters in a lasting focused amazement as I realize that our Sun, which is more than 1,000 times the size of our Earth is orbiting around the Milky Ways' Supermassive Black Hole and even that Massive Black Hole is nothing more than a spec of sand amongst the observable universe and that beyond that we have yet to innerstand. And the Milky Way is no small object in comparison to us – it is around 1,000 light years in distance across and just one light year is equivalent to approximately 9.4605284×10^{15} meters or 5,880,000,000,000 miles or 5.88 trillion miles. Mind = Blown.

- **Cedric Paul Harriott**

This one aspect of our lives that so touches everything that we perceive and need is so crucially important to all that we enjoy on this planet, and yet it could all be ended by the will and whim of the universe or multiverse and whatever mechanisms make it work so.

That's insight.

It is amazing that despite this intense information more people aren't attempting to innerstand this grand reality that exists over our heads and within our bodies. We must not allow ourselves to become content with junk television, junk music, junk food, and junk vibrations – nay – we must be good stewards of knowledge and great conduits of information that we may ever increase the level of intelligence and capability of societies with each passing generation. We must remove the blinding handicap of social inadequacies based upon illogical concepts by illogical minds (racism, sexism, etc.) and we must realize that while we bicker on the minutest of matters, there are heavenly objects doing progressions that affect our minds, our bodies, our lives, and all that we know. If we are to survive and thrive in this galaxy, we must be able to be intelligent enough to realize that we are from the same atomic matter. We must be on an ever diligent and responsible journey into funding the research that really matters in the end. The question that I have for you is: While you read this are my words truly changing your perception? Will you adjust your life according to higher vibrational concepts? Or will this just be a good book you picked up a while ago? The choice is yours, and that's insight.

Perception is Reality

Your perspective becomes your reality.
Thus.
Your perception becomes your reality.
Thus.
Your actions become your reality.
Thus.
Your station in life is found.

- **Cedric Paul Harriott**

There are many levels of consciousness, and no one level is higher than the other, in my perception, that is – rather – they coexist in a multiverse concept (side by side in parallel). There are many things that allow one to move effortlessly through the many different levels of consciousness, and my favorite tool is information (or education, or light). I imparted my quote upon you that "Perspectives change when knowledge is obtained". For me, the matrix is a societal system where many people exists in their own bubbles completely oblivious to the high sciences and when you attain light and obtain knowledge or information you are better able to remove yourself from that system and see from a different innerstanding (level) what is happening within us and around us. It is thence that you receive this next quote that I impart upon you that "When you attain more light you are able to live in the matrix and peer outward beyond its barrier and exists outside of the matrix and peer back into it; residing on both planes whenever one wants to; free thinker." In order to be able to truly reach this level of thought, you must study these seven Masonic liberal arts: Grammar, Rhetoric, Logic, Music, Arithmetic, Geometry, and Astrology. You must also being to realize the high science concepts around you and within you. [*Remember that high science isn't always being a master of a particular realm but generally innerstanding principle concepts of certain sciences that allow you to become more rounded in knowledge. For example, I am not a master of biology or physics but I know that we are made from electrons, neutrons, and protons and that those are made from quarks and gluons and that beyond that we aren't able to innerstand what makes up quarks and gluons. This changes my perception because at every article of matters core they are comprised of the same building blocks as I and of the universe, which blows my mind.]* **When you begin to realize these concepts you being to question, wander, wonder, and you minds synaptic connections thence become hyper active (in an esoteric sense) and this allows you to become able to see the different levels of consciousness and thence pick the ones you wish to visit. This may sound like a crazy person's journey and to a person within the matrix that cannot see outside of its boundaries it most likely is a crazy person's journey. Remember that perception and perspectives create reality and if a person hasn't received light then what they perceive of your journey is their truth. Think of it like this: If Amenhotep lives in the District of Columbia and doesn't travel outside of that area then his vision and innerstanding is slightly limited to the grand scope of that which is past the borders. If Ptah-hotep lives in the District of Columbia but traveled outside of those borders on the regular, became an Astronaut and even traveled to outer space then his perspectives and vision is increased more than Amenhotep. Consciousness is like that, and these levels of consciousness are just like that; I assure you that exploring everything and looking like a crazy person is the best feeling in the world especially when you meet another on that same path. It's like when you hear your favorite song and can't help but to get happy and dance and put it on repeat; the euphoria of attaining knowledge is literally just like that all of the time.**

Essence of the Ending Sublime Conclave:

– We Break From Here Away From the Subtext –

In reflection, I revisit the perception manifested in words that I speak softly unto mine own mind that – We are where we are because of who we are – Such a simple saying should now carry such a profound power in your mind. Now that we have traveled together on this grand journey where we have explored some of the inspirational concepts in the high sciences you should have a newly found respect for that perception.

We exist on this seemingly physical plane of existence because of where we are. Where we are in the universe, our innerstanding of our universe, our place of our universe, where we are in life, where we are in our relationships, where we are as it relates to the Earth itself – these play a critical role in the molding and regulation of our perceptions.

We exist, both individually and globally on this current level of spiritual and conscious awareness because of who we are. Who we are within the solemn inner theater of our minds eye, who we are when no one is watching, who we are when we have an audience, who we are when we are at our most vulnerable – these play a critical role in the molding and regulation of our perceptions.

We exist also within the stations of our lives that we exist within because of who we are. Stations being milestone points; student, mother, father, sibling, traveler. – these are different from the individual qualities that make up your sensibilities and such as they are more of a "place" in life as opposed to a "character trait".

You see, where we are defines our perception of the world within us and the world around us. It molds the many chambers of our minds and governs the conclaves that mold who we are within our minds. Where we are governs our relationship to where we should be going in the same way that an oceans current gently guides you down the path you are flowing down – and in the same way that the headwinds of life can either be utilized to your benefit or to your detriment.

Who you are defines the thoughts, sensibilities, actions, motives, and purpose. Who you are is genetically encoded within your DNA and can be passed down from generation to generation - mutations included as you can either be in proper form or to the detriment of your lineage in the same way that radioactive elements mutate DNA which can be passed down from generation to generation.

I say again that we are where we are because of who we are.

We can either work to the benefit of the common good for all of mankind, or we can work to the detriment of the same. There is a balance of good and evil where one cannot philosophically exist without the other and we as the good stewards of knowledge, the mages of the esoteric arts guild must respect this balance and ensure that we nay drift from its gentle subtext.

Some things are seemingly predefined.

How does a dog automatically know to eat grass when it is feeling sick?

How does a bird immediately know how to chirp upon breaking out of the egg?

How is it that birds are seemingly tuned into the magnetic true north of the planet?

And yet some things are dynamic and open to alteration.

The concepts of when you know better you do better come into effect.

If I practice musical instrument X with dedication and consistency, I will get better.

If I study for examination X and review the appropriate material I will get a better score.

If I stay away from poor food and lifestyle choices then I can mitigate much of the health risks associated with poor food choices and lifestyle choices.

Again I say we are where we are because of who we are.

My closing questions to you are:

Who are you?

And.

Where are you?

End.

www.ingramcontent.com/pod-product-compliance
Lightning Source LLC
Chambersburg PA
CBHW041312110526
44591CB00022B/2893